Hybrid Electric Vehicle

David Agam

Published - 05/23/2015

ISBN-13: 978-1512325966
ISBN-10: 1512325961

Library of Congress Control cataloging -publication data.
Create Space Independent Publishing Platform, North Charleston, SC.

Manufactured in the United States of America, 2015.

Introduction

This book guides the consumer and technician through the Hybrid Electric Vehicle technology complex and as a result, builds professional capacity through that effort to ensure success. This book delivers leading-edge information not available elsewhere that everyone has been begging for, while most books confuse you with insufficient data about this technology.

We live in a mobile and globally connected society. More than ever consumer and technician need to understand the hybrid technology complex. Before you buy or build a hybrid vehicle, you must understand this technology. This book refers to the most recent characteristics and operation of hybrid vehicle systems.

What will you find in the book?

Hybrid vehicle

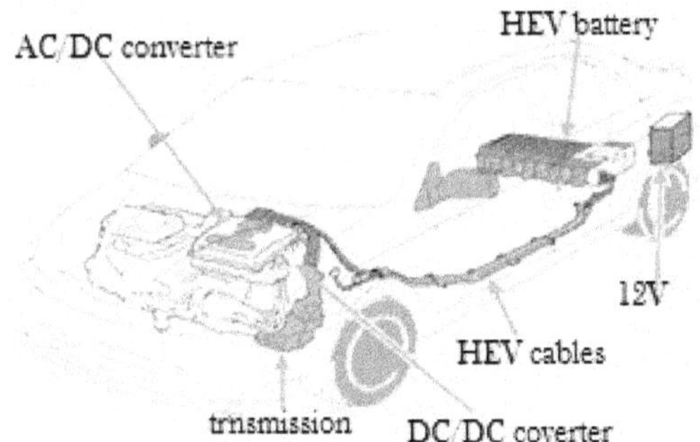

Hybrid electric vehicles merge internal combustion engines (ICE) with electric motors, offering improved fuel efficiency over cars that run on gasoline only. Hybrid model vehicles have no plug like electric vehicles because the batteries recharged from regenerative braking technology, electric motors or gasoline engines. In the past years Hybrid electric vehicles technologies were hardly known, but today, they have come a long way since their popularity increased, and automakers are busy designing and building new models every year.

Hybrid electric vehicles also produce less tailpipe emissions and give much better gas mileage – when today's models get up to 60 MPG.

There is no need to connect the hybrid vehicle to recharge, so a hybrid vehicle needs no access to electricity. However, different levels of hybridization between Hybrid electric vehicles exist in the market. Hybrid models classified by the way they are designed, such as a series or a parallel drivetrain system. The drivetrain comprises all the components needed to transfer power to the wheels. Series hybrids offer various advantages and disadvantages when compared to parallel hybrid designs. Hybrids vehicles can overcome the flaws of internal combustion engine and are free from foreign sources or fuel products.

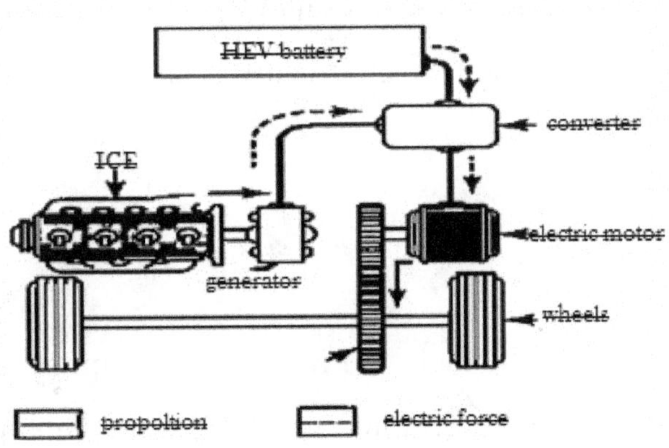

In addition, at low speeds, the electric motor produces greater torque. Both diesel engines and ICE must run at high speed before they could produce great torque. Hybrid electric vehicles related Problems such as expensive components like battery, electric motors, motor controllers, and other electronic components, and sensors. Hybrid electric vehicles optimizing the fuel consumption and the internal combustion engine operation. The characteristics of most common Hybrid electric vehicles, includes the Stop & Go system, the regenerative braking, and the electric drivetrain system.

Parallel drivetrain components include the gasoline engine, electric motor, computer controls and battery. The engine and motor are parallel to one another in design and they both connect to the transmission to thrust the wheels.

The computer controls and coordinates the scheduling and performances of the ICE engine and the electric motors operation. They operate independently or together, depending on the needed force. The battery is, in part, recharged from regenerative braking, a technology that allows the storage of kinetic energy from coasting and braking. The electric motor can also help recharge the battery.

A starter/generator system is still in use, and considered a hybrid vehicle. The idea is to replace the alternator with an electric starter/generator motor as a driving force of the vehicle. A starter/generator system includes a 36 volts battery, and a motor /generator, which is positioned in front of the engine. The engine drives this motor through the front belt.

When the engine is running, the electric motor runs as a generator, which requires a separate system of 36 volts. (42-volt when the battery being charge).

During starting, the electric motor is provided 36 volt for starting the engine through a front belt. The vehicle's conventional starter is used only when the engine is cold. This operating system is used in several types of Hybrid electric vehicles from GM and Saturn vehicle.

A mild hybrid system comprises of features Stop and Go and regenerative braking, but cannot use the electric motor of the vehicle without the help of the ICE. The main role of the electric motor is to allow the shutdown and starting the ICE, according to travel requirements.

Mild hybrid system has the advantage of low cost, but it saves less fuel compare to the Full Hybrid electric vehicles. Mild hybrid system normally uses a 36 volts electric motor, and a 42-volt battery pack.

* * * * *

The medium hybrid systems are equipped with 144 volts battery and provide the Stop-and-Go, a regenerative braking, and auxiliary power. As a mild hybrid system, a hybrid system cannot drive the vehicle by the battery power only.

Models of medium hybrid drivetrain include cars Honda Insight, CIVIC, and Accord. The savings in fuel consumption are between 12-15%.

The Full hybrid drivetrain also called a strong hybrid, use the Stop-and-Go, regenerative braking, and auxiliary power, and the ability to drive the vehicle using the electric motor only. The full hybrid system uses a 200 - 330V battery.

Models of Hybrid electric vehicles with full hybrid system include the Ford and Toyota models. Fuel consumption savings are about 20-25% for the Full hybrid system. Electric motor control system can achieve efficiencies over 90%, compared to 35% or less in the conventional engines. The engine-estimated power is usually much higher than necessary on the road. When travelling at a cruise speed the engine has no overload capability, like electric motors. Therefore, this is why running the engine at idle speed when the vehicle is not moving reduces its effectiveness, compared with higher speed operation.

The components of the series drivetrain include an electric motor, gasoline engine, computer controls, battery and generator. In a series hybrid, only the electric motor can propel the wheels. Rather than operating in parallel, the energy to drivetrain delivered in series, from one power source, the gasoline engine, to another, the electric motor. The gasoline engine turns the generator that powers the electric motor. Batteries recharged from regenerative braking, from the engine, and the generator. The computer decides how much power to the motor comes from the battery or the engine or generator.

The need for a complicated multi-speed transmission and clutch was eliminated in the series hybrids since the electric motor is the one to propel the wheels through the transmission. In series hybrid, the gasoline engines are smaller and more efficient. These design features make series hybrids the ideal car for urban and suburban driving conditions. The smaller, more efficient engine and greater use of electric power helps reduce harmful gas emissions in series hybrids.

The series hybrid requires a larger, more complicated battery and motor to meet its power needs. The larger battery and motor, in addition to the generator often makes the series hybrid more costly than a parallel hybrid. Series hybrids are also not as efficient as much as parallel hybrids for highway driving, because the engine does not directly drives the wheels.

Electrical System

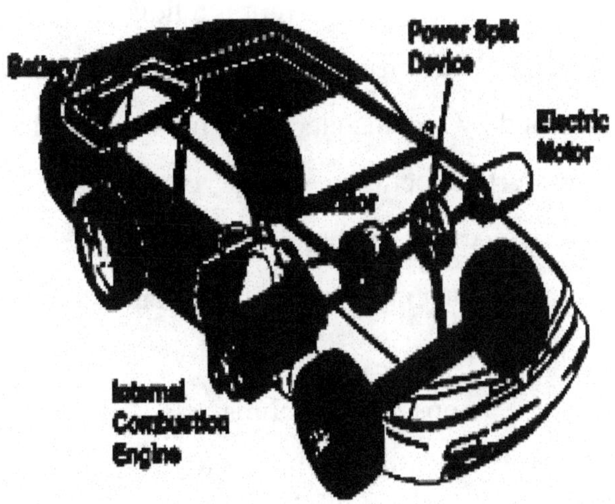

The battery is the only power source for any electrical circuit. However, the current in the circuit is depending on not only resistance but also moves load through the circuit.

Neutrons have no charge, meaning they are neutral. The core of the atom nucleus is positively charged.

An electric current is a flow of electric charge. In electric circuits, this charge is often carried by moving electrons in a by-wire. The connection between voltage, current and the resistance well defined in Ohm's Law.

Current = Voltage: resistance and in terms of units:

Amps = Volts: ohms.

This means that if we multiply the voltage by Current. As the voltage increases, the current will also increases. However, if we multiply the circuit resistance, the current will drop to half its value. Therefore, as the resistance increases, the current will decrease.

Ohm's law states the following formulas:

Voltage - (V), current - (I), resistance - (R),

Where (I) is being the current through the conductor in units of amperes, the V is the potential difference measured across the conductor in units of volts, and R is the resistance of the conductor in units of ohms. More specifically, Ohm's law states that the R in this relation is constant, and independent of the current.

1. I = V/R
2. R = V/I
3. V = I x R

You can apply Ohm's Law by alternating current circuits, including the connection between voltages and current that is more complex than the direct current circuits. AC oscillation exists when a single electron level, there are ' running ' back and forth, without the electrons moving along the applicator more than one move.

The abbreviations AC and DC are refer to an alternating or a direct current or voltage. Direct current (DC) is the unidirectional flow of electric charge. The battery produces direct current. Direct current may flow in a conductor such as by-wire, but can also flow through semiconductors, insulators, or even through a vacuum as in electron or ion beams. The electric charge flows in a constant direction, distinguishing it from alternating current (AC). In direct current, the flow of electric charge is only in one direction. In alternating AC current, the movement of electrons periodically reverses its direction. The usual waveform of an AC power circuit is a sine wave. Certain applications use different waveforms, such as triangular or square waves. Audio and radio signals carried on electrical wiring are examples of alternating current.

Current flowing through a wire heats the wire. The length of a wire affects its resistance, which determines how much current flow in the by-wire and how hot the wire gets.

Magnetic field

A magnet is a material or object that produces a magnetic field. This magnetic field is invisible but is accountable for the most notable property of a magnet: a force that pulls on other Ferro-magnetic materials, such as iron, and attracts or repels other magnets.

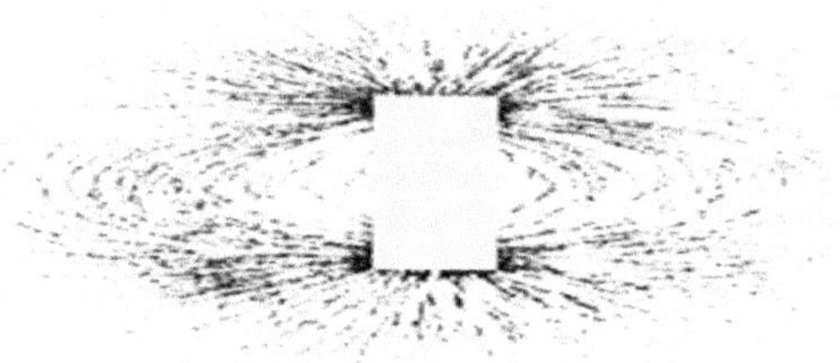

Electrons, moving within the atoms, are creating the magnetic field around a magnet. Although the magnet itself is stationary, its electrons are in constant motion. There are two types of magnetism movement: the spin movement of the electrons, and circular movement around the atom.

Most of the magnetized iron has north and south poles as the magnetic field of the Earth applied. Every magnet has a north pole and both the South Pole.

Magnetic pole has just two poles at both ends are distant. The North magnetic pole cannot exist without the South Pole. Magnetic pole broken in half-and-half will act as a magnet. Broken into two parts again-and will grow four magnets. Even if you continue to break, the magnets will never be able to isolate a single magnet pole. Even when the magnet is one atom thick, it will still have two poles. We can therefore conclude that the atoms themselves are Magnets. Horseshoe magnet is nothing but a magnet bent into a horseshoe with the poles at both ends.

An electro-magnetic field is a physical field produced by an electrically charged object. It affects the behavior of charged objects near the field. The electro-magnetic field extends indefinitely throughout space and describes the electro-magnetic interaction. It is one of the four fundamental forces of nature (the others are gravitation, weak interaction and strong interaction).

Field can be viewed as a combination of electric fields and magnetic fields. Electric fields are produced by stationary charges and the magnetic field by the flowing electricity (currents). How charges and currents interact with the electromagnetic field is described they are characterized by frequency and shaft length.

Equal poles repel each other and different poles attract each other. This rule is similar to rule from the forces between electric charges. Electric charge can be isolated, but not isolated at the magnetic poles. Electrons and protons are entities themselves. These electrons, protons etc. or other units of energy even differentiate themselves from each other. The specific structure or constitution of each such entity is the product of the atom.

There is a vital difference between the magnetic poles for electric charge. In addition to the force, which we call electric power, there is another force resulting from the charges and it called the magnetic force. Electric force and magnetic force are electromagnetic phenomenal.

Magnets exert forces on each other in non-contact. The strength of the magnetic force exerted depends on the distance between them. Magnet hanged by a thread will run as a compass.

If the space surrounding an electric field is in motion, the surrounding space is varying even more. This change is the magnetic field.

The electric field is storage of energy. When we change the flow direction, the magnetic field direction will change as well.

If you insert a piece of iron into a superconducting coil carrying current, the molecules of the conductor will set along the iron poles, which increases the strength of the magnetic field even more, and created an electromagnet.

Relays

Most Hybrid electric vehicles Use common relays so it is imperative to know the theory behind the performance, function and application of the relay.

A conventional vehicle uses about 20 or more relays in a wide range, depending on the applications, and the current in the circuit ratings. A computer or power controllers control most relays.

A relay is an electrically operated switch. Many relays use an electromagnet to operate a switch mechanically, but other operating principles are also used, such as solid-state relays. Relays are used where it is necessary to control a circuit by a low-power signal (with complete electrical isolation between control and controlled circuits), or where several circuits must be controlled by one signal.

A contactor is type of relay that can handle the high power, which require controlling an electric motor or other loads. Some relays control power circuits with no moving parts, instead using a semiconductor device to perform switching. Relays sometimes used to protect electrical circuits from overload or faults.

When the relay receives a command signal (voltage), the coil creates a magnetic field. Magnetic field (high current) closes the circuit and allows the passage of current-high. Relay's circuit is either open or close, without intermediate states.

When the relay receives a command signal (voltage), the coil generates a magnetic field. The magnetic field (strong current) closes the circuit and allows the transfer of a high current. The strength of the magnetic field is higher as the current increases. Most relays have a parallel connection Diodes to protect against current spikes during operation of the relay. There are two types of relays: In one type, the relay closes the circuit when it is activated. In the second type, the relay opens the circuit when activated. In both the relay operates under the same basic principle. The relay can have moisture in it and cause failure during operation. Therefore, it is difficult to diagnose problems caused by relay moisture.

Capacitors use static electricity (electrostatics) rather than chemistry to store energy. Inside a capacitor, there are two metal plates with an insulated material called a dielectric in between them. When charging a capacitor the positive and negative electrical charges build up on the plates and the separation between them, which prevents them coming into contact, is what stores the energy. The dielectric allows a capacitor of a certain size to store more charge at the same voltage.

Capacitors have many advantages over batteries: they weigh less and generally do not contain harmful chemicals or toxic metals, and they can be charged and discharged many times without ever wearing out. However, they have a big drawback too: their basic design prevents them from storing anything like the same amount of electrical energy as batteries.

In an ordinary capacitor, the plates are

separated by a thick dielectric made from something like mica (a ceramic), a thin plastic film, or even air (in something like a capacitor that acts as the tuning dial inside a radio). When the capacitor is charged, positive charges form on one plate and negative charges on the other, creating an electric field between them. The field polarizes the dielectric, so its molecules line up in the opposite direction to the field and reduce its strength. That means the plates can store more charge at any given voltage.

A super capacitor differs from an ordinary capacitor in two important ways: its plates effectively have a much bigger area and the distance between them is much smaller because the separator between them functions in a different way to a conventional dielectric.

Like an ordinary capacitor, a super capacitor has two plates that are separated. The plates are made from metal coated with a porous substance such as powdery, activated charcoal, which gives them a bigger area for storing much more charge. Imagine electricity is water for a moment: where an ordinary capacitor is like a cloth that can mop up only a tiny little spill, a super capacitor’s porous plates make it more like a chunky sponge that can soak up many times more. Porous super capacitor plates are electricity sponges!

In a super capacitor, both plates are soaked in an electrolyte and separated by a very thin insulator. When the plates are charged up, an opposite charge forms on either side of the separator, creating what's called an electric double-layer, maybe just one molecule thick.

This is why super capacitors are often referred to as double-layer capacitors,

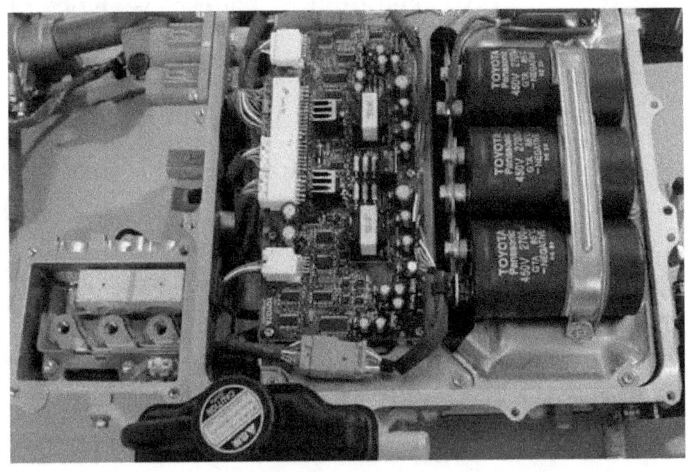

Conventional vehicle

An automotive battery is type of rechargeable battery that supplies electric energy to the vehicle. An automotive battery powers the starter motor, the lights, and the ignition system of a vehicle's engine.

Automotive batteries are led-acid type and are made of six galvanic cells in series to provide a 12-volt system. Each cell provides 2.1 volts for a total of 12.6 volts at full charge. Heavy vehicles, Lead-acid batteries are consisted of plates of lead and separating plates of lead dioxide, which are submerged into an electrolyte solution of about 38% sulfuric acid and 62% water. This causes a chemical reaction that releases electrons, allowing them to flow through conductors to produce electricity.

As the battery discharges, the acid of

the electrolyte reacts with the materials of the plates, changing their surface to lead sulfate. When the battery is recharged, the chemical reaction is reversed: the lead sulfate reforms into lead dioxide and lead. With the plates restored to their original condition, the process now repeated.

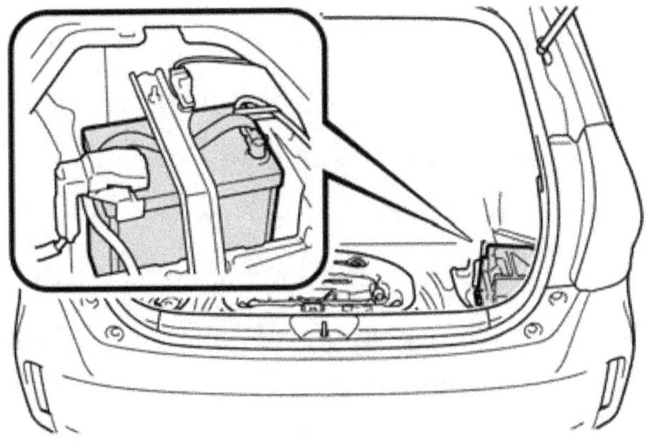

However when the vehicle is driven main electricity provider is the alternator. The main purpose of the battery is to start the engine and provide power when the vehicle is not driven. The battery converts chemical energy into electrical power.

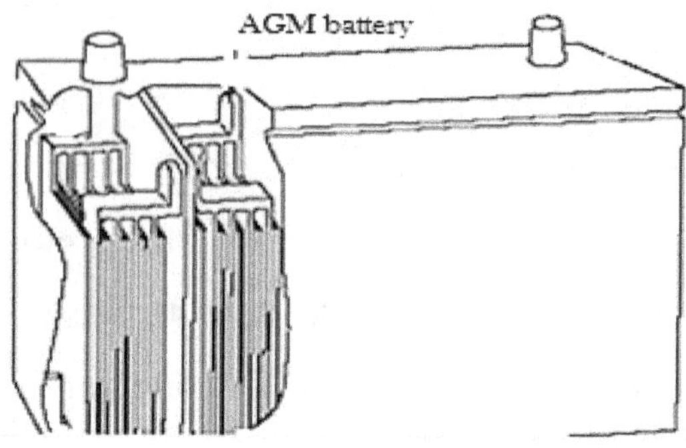

Hybrid electric vehicles also contains a lead-acid 12 Volt battery. This 12 Volt auxiliary battery powers the electrical system the same way as conventional vehicles. As with other conventional vehicles, the auxiliary battery is grounded to the metal chassis of the vehicle. Auxiliary battery is located in the cargo area on the right side in the rear quarter panel. The DC Converter charges the hybrid vehicle battery. The most common battery hybrid electric vehicle lithium battery is AGM Glass Mat.

*****

AGM or Absorbed Glass Mat batteries are the most efficient of the lead-acid design. AGM batteries use a fiberglass separator to keep the electrolyte between the lead plates. This makes AGM style batteries extremely stable in any position, even upside down. These batteries last much longer than a conventional battery and are the superior design for wet-acid type vehicle batteries. AGM batteries also resist freezing longer than conventional batteries.

A lithium battery of AGM considered

the most efficient and sustainable. Is opaque, but is equipped with a pressure relief valve, and can be placed in any position without the risk of leakage of liquid or toxic gas and has a low self-discharge. Battery capacity shows the number of minutes the battery can deliver 25 amps and stay above 10.5 volts.

Reliable method to estimate the battery charge status is checking open circuit voltage using a digital voltmeter (DMM) through the battery terminals.

This test operates headlights for one minute. After turning off the headlights, battery voltage level in a few minutes (battery voltage ceases to rise), then check the voltage reading.

It is also possible to diagnose the condition of the battery by measuring its conductivity. Once determined that the battery is at least 75% charged, so you can check the battery capacity and load.

The load is determined by dividing an ampere-hour (CCA) of the battery at 2.

Another method is to measure the internal resistance of the battery using special test equipment.

All tests must be carried out by the battery manufacturer's instructions. The battery cables must be disconnected before replacing any components.

The ECM computer controls the battery charging of the AGM glass.

High Voltage Battery (HV)

The hybrid technology is speeding up battery technologies to where battery-powered vehicles may become more common. By connecting the batteries in series to each other, they can provide high enough voltages to power electric vehicles.

The battery pack in a hybrid vehicle is made of several cylindrical cells or prismatic cells. These battery packs are often called high-voltage (HV) batteries. There are several types of high voltage (HV) batteries being used or in development.

High-voltage battery composed of cells of 1.2 volts in series. The number of cells and batteries vary between car manufacturers, according to battery voltage requirements.

HV (high voltage) system is equipped with a safety switch master key for maintenance or when opening and closing relays of the high voltage circuit.

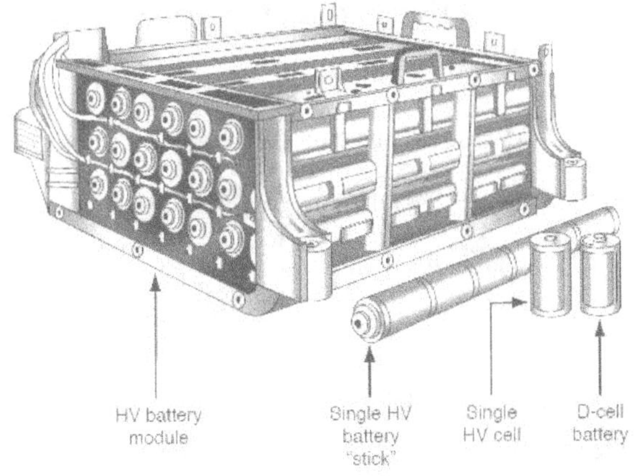

While the vehicle is in motion, the high voltage HV (high voltage) batteries is in a load and unload cycles, while dumps by the electric motor during acceleration and uploaded by recycling braking energy during the downturn.

When the main switch is disconnected service up, it disconnects the high voltage circuit in half the battery voltage and turns off the relays.

For safety reasons, you must always switch off the car to OFF before removing the main service switch.

While charging the high voltage HV (high voltage) battery should be removed from the vehicle.

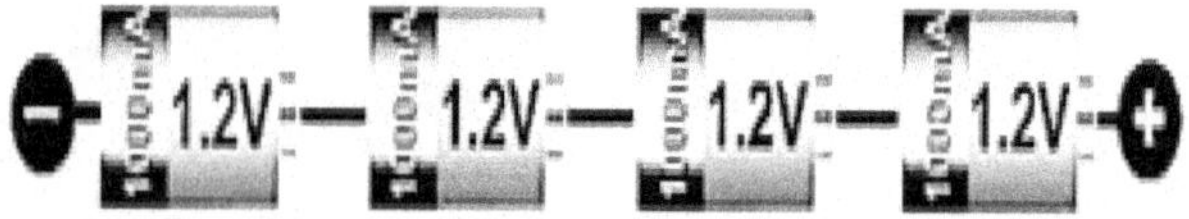

Nickel-Metal-Hydride (NiMH) battery

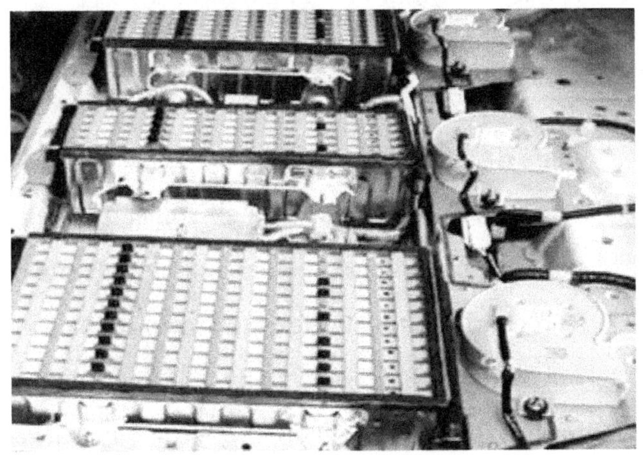

HV (high voltage) battery Nickel-Metal-Hydride is a high energy density. These batteries are damaged faster because they often overcharged, or exposed to excessive heat or high heat environment.

It is vital to measure the battery

required charging time and avoid overcharge. These batteries are charged in a steady current charge and limited time. Overcharging can cause damage to the battery cells and battery explosion.

There are two types of battery designs: cylindrical and prismatic (rectangular shaped).

The NiMH HV (high voltage) battery pack contains sealed batteries that are similar to rechargeable batteries used in some battery operated power tools and other consumer products. In the unlikely event the electrolyte does leak, it can be neutralized with a dilute boric acid solution or vinegar.

All NiMH battery cells produce only 1.2 volts. Each battery has six cells of 1.2 volts, which join the series. The battery has 7.2 volts. To create the battery pack to build a high-voltage battery appropriate number of batteries is connected in series.

For example, to build a pack of 144-volt capable to produce high voltage, connect 20 batteries of 7.2 volts. (144V for the Honda, and 330V for FORD ESCAPE Hybrid vehicle).

The performance of a hybrid electric vehicle depends strongly on the performance of its high-voltage battery pack, which is influenced by temperature. In cold temperatures, batteries perform poorly because of high internal resistance; the vehicle may start slowly.

All electrical energy is converting to heat. Charging the battery fast, cause rapid heating, and will result in poor performance and short life of the battery pack. The HV (high voltage) battery has an electric cooling fan controlled by a battery module. The airflow passes over the battery inside the battery's case and removes the generated heat out of the car.

The HV (high voltage) battery has various temperature sensors in different locations on the battery pack that are sending data to a computer module for controlling battery temperature. These data also used to control the cooling fan operation.

These high voltage batteries have a high self-discharge of about 20% a month and they are vulnerable to discharge state. If the discharge state is below 17-20% of the battery capacity, the battery will stop functioning.

Some manufacturers recommend long period's storage of hybrid electric vehicles, to run it at least 30 minutes every week to prevent damage to the battery and keep it charged.

Hybrid batteries utilize a prismatic module. While these modules are still composed of six individual cells, each module is manufactured together as a single unit, complete with an emergency vent for pressure release during a battery failure event. The second-generation Toyota Prius battery pack contains 28 prismatic modules, and 168 individual cells.

The battery module determines when and how much to charge the battery to maintain the battery charged between 20-60%.

HV NiMH batteries are most common in use in hybrid vehicle due to specific properties, such as energy, life cycle, and safety. Toyota, Honda and Ford, still use NiMH battery packs

Lithium-Ion batteries

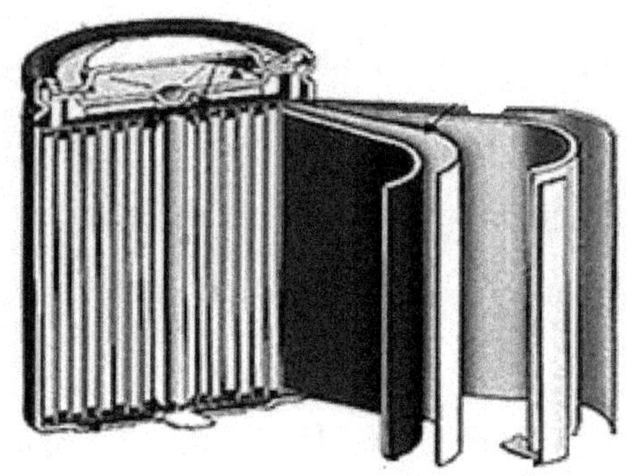

Lithium-Ion Batteries are quickly becoming an efficient energy choice used to power the vehicles of tomorrow. This battery uses the hydrogen absorbing alloy electrode positive.

Systems for hybrid electric vehicles are comprised of several critically important components. In addition to the Lithium-Ion Batteries, the system capability also includes High storage capacity is two to three times more storage capacity of Ni-MH batteries.

Lithium-ion Battery energy have the highest volumetric energy. The rated voltage of the cell-ion is 3.6 volts.

This battery is hardly affected by the 'memory effect' of the storage capacity while in use.

These batteries will soon replace the nickel metal hydride (NiMH), most often used in Hybrid electric vehicles.

Though lithium-ion (Li-ion) batteries have been expensive in the past, they becoming less expensive to manufacture. In the end, as more and more lithium-ion cells are created, automakers can lower the cost of hybrid cars until the prices are closer to that of conventional ICE-powered cars.

The biggest advantage in lithium-ion technology is its ability to be rapidly recharged, which is exactly what is required in hybrid driving. The power stored in the batteries is easily tapped and readily available, and when technology like regenerative braking is factored into the equation, the cells can be recharged rather quickly.

The latest generation of hybrids, like the Chevy Volt, uses Li-ion battery cells. The main component variations in lithium-ion positive electrode, lithium oxide is-cobalt and carbon negative electrode. The battery-ion is so named because the lithium-ion moves back and forth between positive and negative electrode.

To prevent damage to the battery and for safety, the pressure relief valve opens if the internal gas pressure rises above a predetermined point.

Charging and discharging batteries is a chemical reaction, but Li-ion is claimed to be the exception. Battery scientists talk about energies flowing in and out of the battery as part of ion movement between anode and cathode. This claim carries merits but if the scientists were totally right, the battery would live forever. Scientists blame capacity fade on ions being trapped, but as with all battery systems, internal corrosion still plays a role.

The Li ion charger is a voltage-limiting device that is similar to the lead acid system. While lead acid offers some flexibility in terms of voltage cut off, manufacturers of Li ion cells are very strict on the correct setting because Li-ion cannot accept overcharge. The so-called miracle charger that promises to prolong battery life and gain extra capacity does not exist. Li-ion is a "clean" system and only takes what it can absorb.

Lithium-iron-phosphate

The lithium-iron-phosphate battery is a type of rechargeable battery, specifically a lithium-ion battery, which uses lithium-iron-phosphate. It combines a high level of safety in use and in storage, with high capacity storage and supply of energy. By adding metal powders, electron conductivity increased eightfold, or 100 million times. The rechargeable battery will drive a range of 120-160 miles before recharging again. Charging time is 6-8 hours on a power voltage network.

Recharging the battery in the vehicle is possible, with the possibility of quickly replacing the battery in the appropriate service stations.

Zinc-air battery

Zinc-air is a high-energy, high-power fuel cell technology that is safe and environmentally friendly. Zinc–air batteries have some properties of fuel cells as well as batteries: the zinc is the fuel, the reaction rate can be controlled by varying the airflow, and oxidized zinc/electrolyte paste can be replaced with fresh paste. The zinc battery-air can be loaded quickly because a full charge will be achieved by replacing the zinc electrodes. Charging the battery is done by replacing the negative electrode. Negative electrode is biodegradable because during battery discharge cycles and cannot be reuse. The discharged zinc-air module removed from the vehicle is "refueled" or mechanically recharged by exchanging spent "cassettes" with fresh cassettes. A refueling machine that returns the zinc-air modules to service accomplishes this.

The depleted cassettes are electro-chemically recharged and mechanically recycled external to the battery. Regeneration of the cassettes will take place at centralized facilities serving regional networks of refueling stations. In this way, the zinc anode recharging/recycling facility would assume a parallel role in a zinc-air based transportation system to that held by oil refineries in today's fuel distribution system, without the negative impacts.

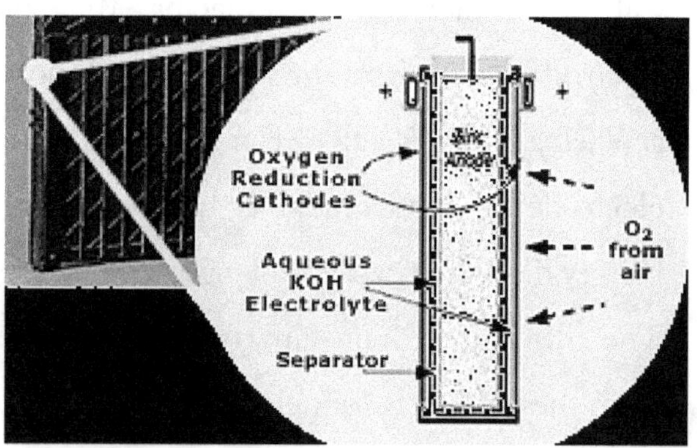

Zinc-air batteries have two major advantages. During discharge, a mass of zinc particles forms a porous anode, which is saturated with an electrolyte. Oxygen from the air reacts at the cathode and forms hydroxyl ions, which migrate into the zinc paste and form zincate, releasing electrons to travel to the cathode. The zincate decays into zinc oxide and water returns to the electrolyte. Hydroxyl mixed with water from the anode is recycled at the cathode, but the water is not consumed.

The reactions produce a theoretical 1.65 volts, but this is reduced to 1.35-1.4V in available cells.

The battery control module determines when and how much to charge the battery to maintain charging mode between 20 -60%.

Battery Control Module

The battery management system (BMS) is a critical component of electric and hybrid electric vehicles. The purpose of the BMS is to guarantee safe and reliable battery operation. To maintain the safety and reliability of the battery, state monitoring and evaluation, charge control, and cell balancing are functionalities that implemented in BMS.

The purpose of the battery control monitors is constantly computing the state of charge for the HV (high voltage) battery in Hybrid electric vehicles. It then sends this information to the battery control unit, which determines whether to charge or discharge the high voltage battery.

Always follow the manufacturer's recommended service procedures when inspecting and replacing the battery control module

If the battery temperature sensor is lower than normal the battery control module interprets this as an open circuit.

If the temperature sensor show higher than normal battery control module interprets this as shorted circuit.

Jump-starting the battery

Jump-starting off the 12-volt battery is possible if the engine fails to start. It is done by using battery of another vehicle. Note that the trunk is equipped with an electric lock latch cannot be opened if the battery is depleted.

If the vehicle is disabled, but can be pushed, the easiest way is to move the gear lever into neutral manually (N) and push the car along the road, or out of the garage when the vehicle service.

It is not permitted to push the vehicle over four miles per hour.

Electric Motors

Brushless Motor

The brushless motor, unlike the DC brushed motor, has the permanent magnets glued on the rotor. It has usually four magnets around the perimeter. The stator of the motor is composed by the electromagnets, placed in a cross pattern with 90o angle between them. The major advantage of the brushless motors is because the rotor carries only the permanent magnets, it needs of NO power at all. This feature gives the brushless motor great increment in reliability as the brushes wear off very fast. Moreover, brushless motors are more silent and more efficient in terms of power consumption.

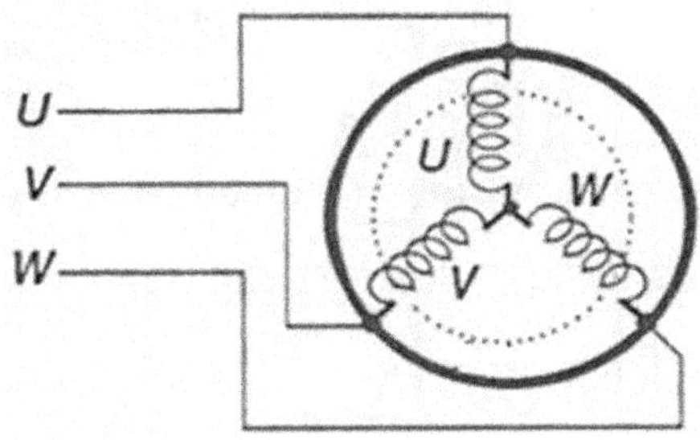

Brushless motors have no commutator or brushes. There are several ways to find out the rotor position. Sometimes they use rotary encoders along with their controllers and they know exactly the angle that the rotor is. Others use pairs of Hall sensors while most of them use just one Hall sensor. The Hall sensor is placed in an appropriate position. It can sense if in front of it is the North or the South Pole. The Hall sensor will then transmit this signal to the controller of the motor. The controller will then switch on or off the appropriate coils needed to provide the torque.

The electric motor structure is similar to that of an electric generator. It converts electrical energy into mechanical energy by producing torque to drive the vehicles.

The electric motor produces maximum torque at low speed, a perfect condition for use as a power source to drive the car from a Stop. Thus, a typical electric motor used in hybrid electric vehicle provides constant torque at low speed, usually between zero and 1500 revolutions per minute, and provides constant driving force above 1500 RPM.

Brushless motors need to have a Hall-effect sensor to measure accurately the angle rotation of the rotor contact with the voltage controller. Most engines have sensors Hall-effect.

Horsepower (HP) is a unit of measurement of power (the rate at which the work is done). The most common horsepower–especially for electrical power is 1 HP = 746 watts. The power of the engine-Electric expressed in kilowatts.

Torque is determined by the strength of the magnetic fields. Magnetic field strength is proportional to the number and thickness of the closely spaced by-wire wraps of the electric motor.

In most Hybrid electric vehicles, there are ten electric motors. These include:

- One and sometimes two electric motors, used to drive the vehicle.
- Auxiliary motor to pump coolant during Stop and Go, to keep the vehicle heat (Honda/Toyota).
- An electric power steering system motor (EPS).
- Electric Air Conditioning compressor
- A transmission cooler electric motor.
- HV (high voltage) battery-motor, cooling fan, (all).

- Electro hydraulic power steering motor (General Motors).
- CVT Parking motor,
- Rear Differential (Lexus, Highlander).

Snubbers

Snubbers are frequently used in electrical systems with an inductive load where the sudden interruption of current flow leads to a sharp rise in voltage across the current switching device, in accordance with Faraday's law. This transient can be a source of electromagnetic interference (EMI) in other circuits.

Additionally, if the voltage generated across the device is beyond what the device is intended to tolerate, it may damage or destroy it. The snubber provides a short-term alternative current path around the current switching device so that the inductive element may be discharged more safely and quietly. While current switching is everywhere, Snubbers will generally only be required where a major current path is switched, such as in power supplies.

IGBT

The insulated-gate bipolar transistor (IGBT) is a three-terminal power semiconductor device primarily used as an electronic switch, which combine high efficiency and fast switching. The insulated gate bipolar transistor is a three terminal, trans-conductance device that combines an insulated gate.

The IGBT is ideal as a semiconductor-switching device.

The IGBTs mainly use in power electronics applications, such as inverters, converters and power supplies, were the demand for the solid-state switching device are not fully met.

The main advantages of using the Insulated Gate Bipolar Transistor over other types of transistor devices are its high voltage capability, low resistance relatively fast switching speeds. The Power Control Unit has six IGBTs transistors.

Resolvers

The resolver is an electromagnetic induction type angular sensor. The rotor consists of laminated silicon steel, and it has no coil. It is considered an analog device, and has a digital counterpart, the rotary (or pulse) encoder.

The most common type of resolver is the brushless transmitter resolver. On the outside, this type of resolver may look like a small electrical motor having a stator and rotor. On the inside, the configuration of the by-wire windings makes it different. The stator portion of the resolver houses three windings: an exciter winding and two two-phase windings (usually labeled "x" and "y"). The exciter winding is located on the top; it is in fact a coil. This transformer induces current in the rotor without a direct electrical connection, and there is no need for brushes. The other windings are located at the lowest part of the transformer. They are configured 90 degrees from each other. The rotor houses a coil, which is the secondary winding of the turning transformer, and a separate primary winding in a lamination, exciting the two two-phase windings on the stator.

In the Toyota Hybrid electric vehicles, the resolver in the electric motor is an extremely reliable and compact sensor that precisely detects the magnetic pole position, for ensuring the efficient control of MG1 and MG2.

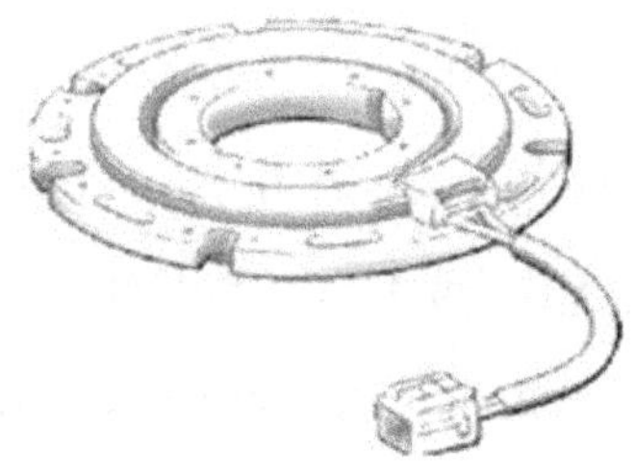

Ultra-Capacitors

An Ultra-Capacitor, also called a super capacitor, is an electrochemical capacitor with a higher energy density than normal capacitors, which potentially makes them a better fit for Hybrid electric vehicles.

Ultra-capacitors store electricity by physically separating positive and negative charges. A major advantage of ultra-capacitors is their ability to capture electricity from regenerative braking and provide that electricity to power a car's acceleration. Ultra-capacitors not only charge more quickly than batteries, they also release energy more quickly.

A drawback to their use is the technology's inability to store as much energy as a battery. If that was the case, Ultra capacitors could be more widely used with smaller batteries to power hybrid cars.

The energy stored in a capacitor can be recycled when discharged. Load capacitor is outside negative charge positive polarity to negative when the condenser is located inside the circle is closed, or connected to a power source like a battery, there is a flow of electricity to up the voltage capacitor voltage becomes identical to the original.

When the ignition turned off, and the battery disconnected, car manufacturer's advice you must wait 5 - 20 minutes to empty the Ultra capacitors. You should perform Ultra capacitors test with protective gloves to prevent electric shock until proven that the capacitors are empty.

Converter cooling system

Power flow through the electronic control units in hybrid vehicle tends to create much heat. Toyota's car manufacturers, Ford, and GM use liquid cooling to control the temperature of these electronic units.

DC-DC Converters

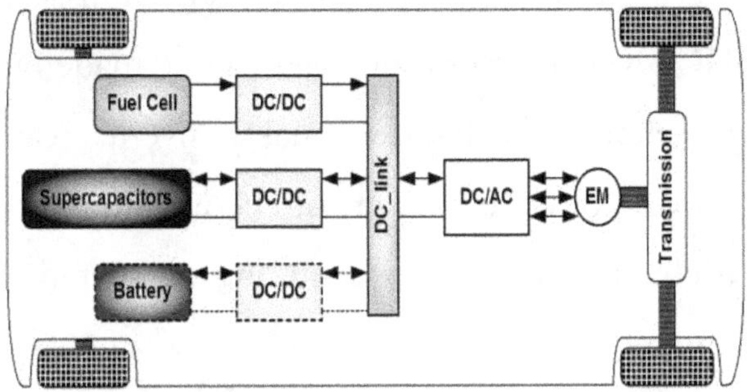

A DC-to-DC converter is a category of power converters and it is an electric circuit, which converts a source of direct current (DC) from one voltage level to another, by storing the input energy temporarily and then releasing that energy to the output at a different voltage. The storage may be in either magnetic field storage components (inductors, transformers) or electric field storage components (capacitors). At least one DC/DC converter is necessary to interface the Battery or the Super capacitors'

module to the DC-link.

DC/DC converters can be designed to transfer power in only one direction, from the input to the output. However, almost all DC/DC converter topologies can be made bi-directional. A bi-directional converter can move power in either direction, which is useful in applications requiring regenerative braking.

The amount of power flow between the input and the output can be controlled by adjusting the duty cycle (ratio of on/off time of the switch). Usually, this is done to control the output voltage, the output current, or to maintain a constant power.

One example of a DC-DC converter circuit is a computer circuit that converts a voltage of 14-volt voltage of 5V, which is used to feed the engine sensors with an ECM computer system.

Hybrid electric vehicles use DC-DC

converters to supply higher or lower voltage according to the requirements of the vehicle, to charge the 12 volts battery, and to operate accessories and lighting systems, etc.

AC-DC Converter-Inverter

AC/DC converter 1

AC converter (Inverter) is an electrical device that converts direct current (DC) current (AC). The main component that makes the current is the converter controls the operation of six IGBT transistors.

Typical efficiency of these converters is 95% or more.

Inverters and converters combined into one unit manage the power and recharging circuits in hybrids and electric vehicles. An inverter is an electrical device that converts electricity derived from a DC (Direct Current) source to AC (Alternating Current) that can be used to drive AC utilization. The theory of operation is relatively simple.

DC power, from a hybrid battery for example, is fed to the primary winding in a transformer within the inverter housing. Through an electronic switch (generally a set of semiconductor transistors), the direction of the flow of current is continuously and regular broken (the electrical charge travels into the primary winding, then abruptly reverses and flows back out). The in/out flow of electricity produces AC current in the transformer's secondary winding circuit. Ultimately, this induced alternating current electricity flows into--and produces power in--an AC load (for example an electric vehicle's (EV) electric traction motor). A rectifier is a similar device to an inverter except that it does the opposite, converting AC power to DC power.

The inverter is more properly called a voltage converter; this electrical device changes the voltage (AC or DC) of an electrical power source. There are two types of voltage converters: step up (which increases voltage) and step down (which decreases voltage).

The most common use of an inverter is to a take relatively low voltage source and step-it-up to high voltage for heavy-duty tasks in a high power load, but they can be reversed to reduce voltage for a light load source.

An inverter/converter is one single unit that houses both an inverter and a converter. These devices are used in hybrids vehicles to manage their electric drive systems. Along with a built-in charge controller, the inverter/converter supplies current to the battery pack for recharging during regenerative braking, as well as provide electricity to the motor/generator for vehicle propulsion.

Both hybrids and EV vehicles use relatively low voltage DC batteries (about 210V) to keep the physical size (and cargo space consumed) down, but they also generally use highly efficient, and high voltage (about 650-V) AC motor/generators. The inverter/converter unit composes how these different voltages and currents works.

Because of the use of transformers and semiconductors (and the accompanying resistance encountered), enormous amounts of heat are emitted by these devices. Adequate cooling and ventilation are keeping the components operational, and as such, inverter/converter installations in Hybrid electric vehicles have, their own dedicated cooling system (complete with pumps and radiators) that are entirely independent from the engine's cooling system.

Gasoline Engine

– ICE

The hybrid car has an internal combustion engine much like the one you will find on most cars. However, the engine on a hybrid is smaller and uses advanced technologies to reduce emissions and increase efficiency.

Most Hybrid electric vehicles use engines with four performances Action: suction, compression and exhaust action, starting with the starter motor, which rotates the engine two pulses per revolution of the crankshaft. Each cylinder is repeatedly pulsed.

The engine converts some of the energy of the fuel for driving power. This power is used to move the vehicle.

Engines used in hybrid electric vehicles are different from those used in conventional Hybrid electric vehicles.

These differences can include:

1. The engine capacity is smaller than

comparable vehicles of the same size weight.

2. The use of 'Atkinson stroke' to increase the engine's efficiency.
3. The use of conventional Starter.
4. The car manufacturers use an index spark plug so that the open gap of the spark so to point at the intake valve to obtain maximum efficiency.
5. The ECM computer is used to disable or eliminate unwanted engine vibration controls the motor mounts.
6. The use of lower viscosity engine-oil, such as 0W -20.

Atkinson cycle

The Atkinson cycle is designed to provide efficiency at the expense of power density and is used in some modern hybrid electric applications. Similarly, the cycle was design to minimize the use of fuel during the intake stroke yet exploit the part of the cycle that generates power.

to achieve significant improvements in thermal efficiency, the Atkinson Cycle engines have been lavished with numerous innovations, including the reshaped intake port, increased compression ratio, idling-stop, and Variable Valve Timing intelligent Electric (VVT-iE).

During the intake stroke, the intake valve opens before the piston starts its downward travel, but instead of closing the valve as the compression stroke begins, the intake valve is held open even as the compression starts. The intake valve finally closes very late in the compression stroke. This delayed closing of the valve allows air in the cylinder to be pushed back into the engine intake manifold, so there is very little vacuum created in the manifold. This reduces engine-pumping losses, or the restriction caused by a vacuum as the piston goes through the intake stroke. The rest of the cycle continues with compression, power and exhaust the same as the conventional cycle four-stroke design.

Modification of the piston's connecting rod design allows the Atkinson cycle, allows the piston to perform all four-stroke in a single crankshaft round, and allows any stroke to be different in length. The length of the inlet and exhaust strokes is longer than the compression stroke.

One feature of the Atkinson cycle and remains in use today is the intake valve is held open longer than usual to allow a counter flow into the intake manifold, to achieve a better fuel consumption.

Hybrid vehicle can take advantage of the Atkinson Cycle to drive the vehicle by an electric motor at low speeds.

Variable Camshaft Timing

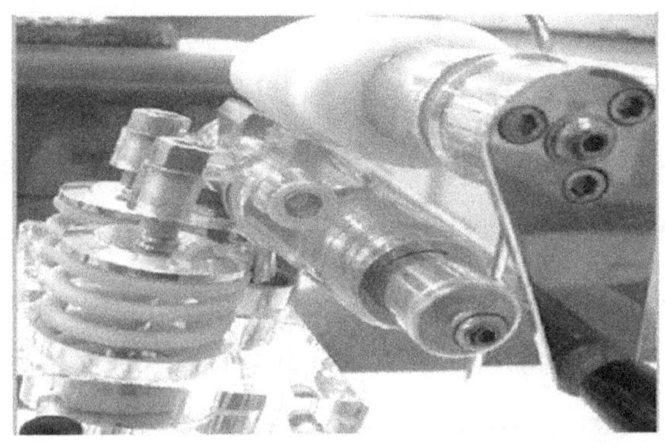

Variable Camshaft Timing (VCT) is a technology developed by Ford. It allows an optimum engine performance, reduced emissions, and increased fuel efficiency compared to engines with fixed camshafts. It uses electronically controlled hydraulic valves that direct high-pressure engine oil into the camshaft Phaser cavity. These oil control solenoids are bolted into the cylinder heads towards the front of the engine near the camshaft Phasers. The electronic control module (ECM) transmits a signal to the solenoids to move a valve spool that regulates the flow of oil to the Phaser cavity. The Phaser cavity changes the valve timing by rotating the camshaft slightly from its initial orientation, which results in the camshaft timing being advanced or retarded. The ECM adjusts the camshaft timing depending on factors such as engine load and RPM.

Variable timing camshaft engine used in General Motors and engines 4 and 6 - cylinder engines of BMW and Nissan.

The use of rollers for lifting the valve reduces friction in the valve.

Variable valve timing

Variable valve timing (VVT) is the process of altering the timing of a valve lift event, and is often used to improve performance, fuel economy or emissions. It is increasingly being used in combination with variable valve lift systems. This can be achieved, ranging from mechanical devices to electro-hydraulic and Camless systems in many ways. Increasingly strict emissions regulations are causing [citation needed] many automotive manufacturers to use VVT systems.

Most four-stroke piston engines today employ one or more camshafts to operate poppet valves. The lobes on the camshafts operate cam followers, which in turn open the poppet valves. A Camless (or, free valve engine) uses electro-magnetic, hydraulic, or pneumatic actuators to open the poppet valves instead. Actuators can be used to both open and close the valves, or an actuator opens the valve while a spring closes it.

Camless engine

The camshaft alone decreases the mechanical efficiency of an engine since the friction of the cams and the load of a timing belt increase the irreversibility's of an engine.
A Camless engine uses electro-magnetic, hydraulic, or pneumatic actuators to open the valves. Actuators open and close the valves, or open the valve while a spring closes it. With the Camless engine, where lift and valve timing can be adjusted freely from valve to valve and from cycle to cycle. It also allows multiple lift events per cycle and, indeed, no events per cycle—switching off the cylinder entirely.
Camless engines are not without their problems though. Common problems include high power consumption, accuracy at

high speed, temperature sensitivity, weight and packaging issues, high noise, high cost, and unsafe operation in case of electrical glitches occurs. Although some Camless systems are commercially available, they are not in production in car engines.

Computers became better after transforming from analog-to-digital. However, several conditions are required to make thing digital. The first condition for being digitally control, is to have an 'On' and 'Off' modes. The second is that the mechanical part can only require energy to change between the on and off mode. The mechanical component cannot use any energy to hold itself in either mode. This allows the components to be much more efficient and extremely fast.

This technology enables reliable and cost-effective variable valve timing in Camless internal combustion engines. The first generation valve design supports operation over 6500 rpm. Variable valve timing technology has demonstrated a fuel efficiency improvement of up to 20%.

Currently, variable valve timing mechanisms are either too costly to implement on conventional vehicles, or far less effective and robust than desired.

Valves of this type can be applied to a wide variety of internal combustion engines. An electromechanical valve actuator eliminates the engine components required for a typical camshaft drive, therefor decreasing manufacturing costs.

The Variable Valve Timing and Lift Electronic Control (VTEC) engine is the latest development designed by Honda to enhance the efficiency of an engine. This system changes cam profiles from the use of synchronized pins controlled and powered by electro-hydraulic devices. The theory behind this technology is to increase the amount of air supplied to the engine as the engine speed is increased. Camless VVT allow the engine to experience maximum engine performance and fuel efficiency at every engine RPM while following principles of the VTEC.

The Camless VVT is electronically operated by an ECM and logic a modulators, which controls electro-mechanical rotary solenoid actuators.

The rotary solenoids offered meet the specifications needed for the required amount of force and speed needed from a solenoid in order for this system to be effective. Likewise, each of these solenoids could connect to the intake and exhaust valves. The ECM supplies and directly controls the precise amount of air flowing through the cylinders for every engine speed.

Depending on the load and engine speed, the ECM will process the information of load and speed given and use a pulse width modulator activate the solenoid valves. This modulator will generate square waves identical to the timing profile for the most efficient engine at that particular speed and load combination.

Active Fuel Management

In some engines, some of cylinder could be disabled during a low load to improve fuel consumption.

When the computer determines that it can deactivate cylinders, it directs oil through special oil passage, which pushes a pin to disable the valve lifter. With the lifter unable to lift the valve, it keeps the valve in the closed position. An electronic system monitors the oil flow by an electric solenoid that used to activate or deactivate the cylinder.

Some Toyota engines have variable valve timing- intelligent (VVT-I), which uses sophisticated computer control system variable valve timing and ensures constant valve overlapping.

The result is higher productivity of torque at any engine speed, improving fuel efficiency and reducing carbon monoxide, hydrocarbon, and nitrogen oxide emissions.

Vehicles equipped with VVT-I systems offer better performance than vehicles with standard engine of the same size.

Engine vibration control

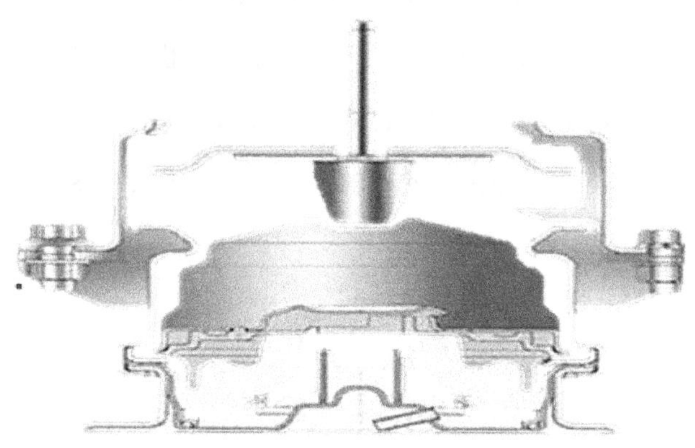

When the engine equipped with cylinder deactivation system, and natural imbalance occurs, it produces vibration and tapping sounds. Those engine mounts are equipped with a sensor that sends signals to the ECM. The ECM counteracts the engine vibration by electronically controlling the engine mounts themselves against the engine vibration.

In addition, the boom sounds that heard are reduced by an automated system of Active noise-control, sending high decibels sounds through the radio speakers, to counteract the engine noises. This system stays active even when the radio is not playing.

Ignition system

The ignition system contains components and wiring that is required to create and control the high voltage (up to 40,000 volts or more). The ignition coils are transformer devices in automobile ignition systems, which produce the high voltage necessary to fire the spark plugs of the engines.

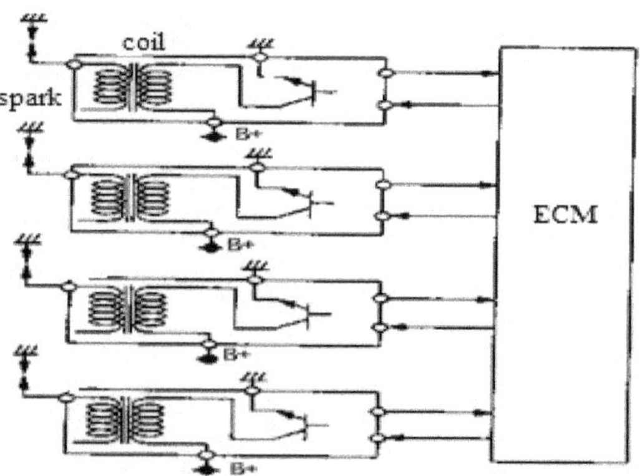

The primary winding in the ignition coil has few loops and less resistance.

Applying the battery voltage to this coil causes a sizable DC current to flow even though it has a significant inductance, which slows the increase of that current to its maximum value. The secondary coil has a much larger number of turns and therefore acts as a step-up transformer. Instead of operating on AC voltages, this coil is designed to produce a large voltage spike when the current in the primary coil is interrupted. Since the induced secondary voltage is proportional to the rate of change of the magnetic field through it, opening a switch quickly in the primary circuit to drop the current to zero will generate a large voltage in the secondary coil according to Faraday's Law. The large voltage causes a spark across the gap of the spark plug to ignite the fuel mixture.

In Hybrid vehicle there is a coil-on-plug for each cylinder. The coil-on-plug connected directly to the spark plug without ignition cables. The engine ECM controls the coil-on-plug systems.

The ECM controls the ignition timing. Most Hybrid electric vehicles use this ignition system

Index Plugs

The Honda Insight hybrid uses an index -lighter to prevent three and four cylinders. Plugs Index-means that the side of the spark plug electrode is directed contrary to the suction valve. With Conventional plugs, side electrode may block or not to block the electric spark.

Honda has four different spark plugs for this engine, called A, B, C or D.

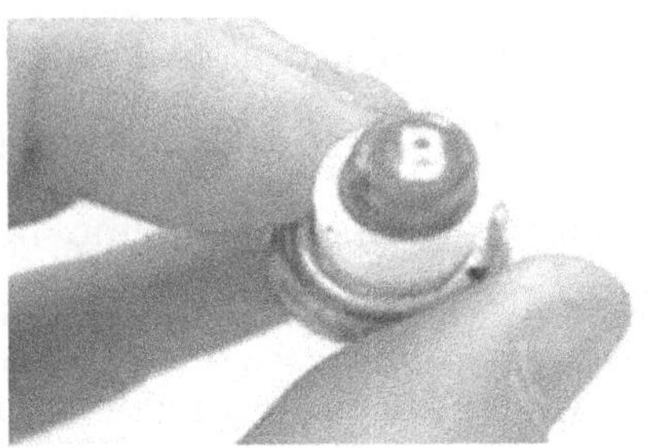

Ion sensing Ignition system

In the ion-sensing ignition, spark plug itself becomes a sensor. The ECM apply 100 to 400V through the spark plug electrodes gap, after the spark occurred to sense the burning gases, called plasma inside the cylinder.

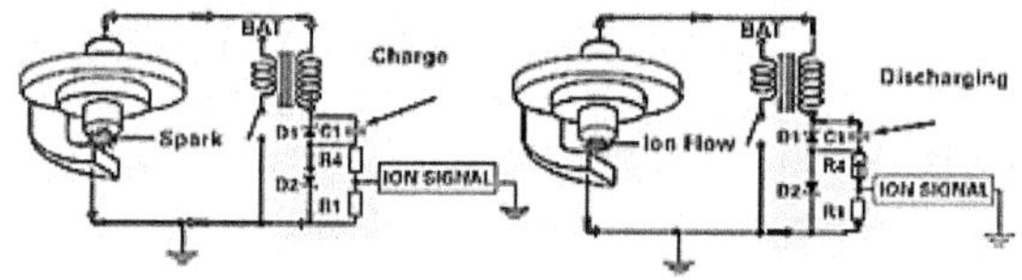

Discharge voltage of the ignition coil is (10 to 15 KW) is electrically insulated from the ion sensing circuit. The flame become ionized and transfers small amount of electricity, which can be accurately measured through the spark plug gap. The reason of doing that is, in ion sensing systems the ignition system operates just like conventional coil-on-plug system but the engine does not need to be equipped with camshaft sensor or knock sensor, because these two functions are achieved by electronic means in the ignition control circuits.

Cooling and heating – Hybrid

The purpose of the engine cooling system is to bring the engine to an optimized temperature as quickly as possible and then maintain that temperature under all driving conditions.

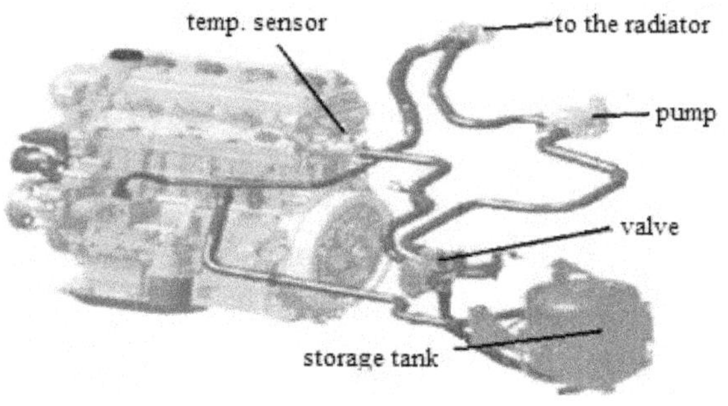

The hybrid system operate an electric auxiliary pump while the engine is stopped (not running, in a Stop and Go condition), and allows the coolant to continue flowing in the cooling system. Some vehicles have a special five-gallon water tank that has the same characteristic of a thermos. The water storage tank can keep the temperature for about three days. The water is transferred to the tank by an electric pump, located nearby. When the engine is turned-off, the water valve does not allow water to pass through in a storage mode. The water valve controls the flow of water from the tank storage, and the heating system. It is controlled by the vehicle BCM. The water valve has a valve position sensor, and an electric motor that operates the valve. The preheating status is activated before the driver starts the engine.

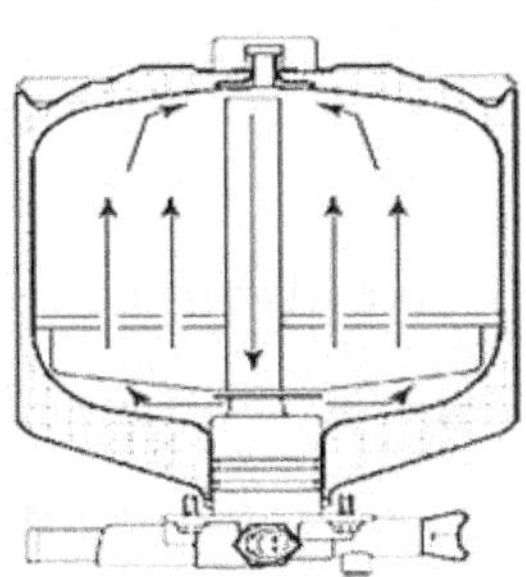

When the engine is cold, the water valve opens to allow the hot water storage tank to move and flow coolant to the cooling system to provide heat start-up heat to the radiator and the heating system. When the engine is stopped, the storage tank filled again by the pump, and then closes the water valve.

This stage is known as the storage stage mode (the ignition switch is off).

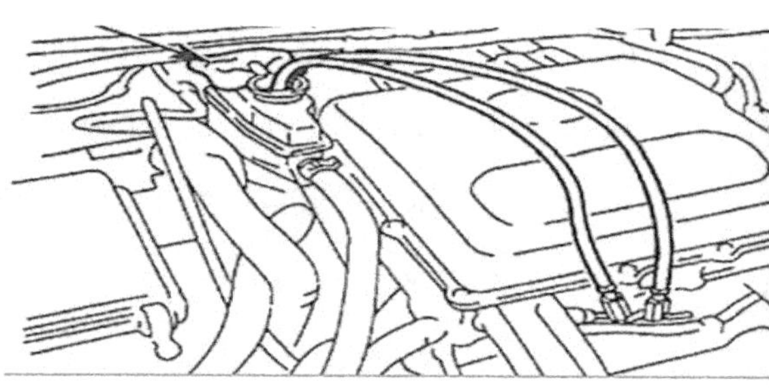

Most of the electronic components in a vehicle will operate with a better efficiency as the temperature reduced, but they will suffer a permanent damage if they are constantly subject to higher than normal heat. All Hybrid electric vehicles have cooling system and control units, and some use an electric fan to turn the excess heat away from these components.

The cooling system of an electric drive motor of the hybrid vehicle has much in common with conventional cooling systems.

Those cooling system may include an electric motor and a DC-DC converter. The cooling pipes connected to the transmission to cool the electric motor.

Some GM hybrid cars have a separate cooling system that contains its own liquid cooling system, including a heat exchanger, electric water pump, and a radiator. Ford ESCAPE Toyota PRIUS cooling system are similar in design.

The cooling of the high voltage

inverter removes heat from the transmission oil with the same coolant used to cool the electric motors.

Electronic Control unit-

ECM

The ECM is a simple computer that runs software programs. Stored in the main memory chip (PROM) is the software that tells the computer, among other things, how much fuel and spark to deliver to the engine for a given set of operating conditions to ensure optimal engine performance. It does this by reading values from a multitude of sensors within the engine bay, interpreting the data using lookup tables, and adjusting the engine actuators accordingly.

The ECM collects data from sensors, calculates and controls the amount of fuel injected by the injectors and ignition timing, so to control of Air/Fuel ratio.

The data from the sensor and the driver commands are computed according to their importance, based on a number of parameters. The computer checks the sensors' data input and compares them into a predetermine table. If the sensor input is beyond range, for the ECM it is considered a faulty sensor.

The ECM will report to the driver by illuminating the check engine light on the dashboard and keep a freeze mode data and the DTC codes responsible for this failure so a technician could check and fix the problem.

DTC Codes only reports a circuit where a problem may exist, but where exactly in the circuit the problem occurred, the technician must resolve. Most faults that are not air pollution parameter does not cause the check engine light illumination, but saved by the ECM computer.

When the driver key on the ignition, the ECM loads the information off the PROM chip and performs a quick, self-diagnostic test. Some later ECMs also pulse the injectors during this period to aid in what is known as "quick start" (obviously helping the engine startup more quickly). Also during this period, the fuel pump is activated for 2 seconds to “prime” the system. Next, the ECM waits until it receives reference pulses from the crankshaft sensor.

Once it receives reference pulses indicating the engine is turning over, it will use air/fuel ratio values stored in the cranking fuel pulse tables (in the PROM programming) based on the coolant temperature to determine how much pulse width it needs to command the fuel injectors open.

If for some reason the engine should become flooded with fuel, provisions built in the computer help to clear the cylinders. If during cranking of the engine you depress the throttle more than 80%, the ECM will enter “clear flood mode”.

In this mode, the ECM commands a very lean Air/Fuel ratio. The ECM will stay in clear flood mode as long as the throttle is 80% or higher and less than 600RPM.

As soon as throttle position falls below the 80% threshold or go above 600RPM, the ECM disables the clear flood mode and calculates fuel delivery based on coolant temperature and other factors it normally uses.

In run mode, the ECM operates in two conditions commonly referred to as OPEN and CLOSED LOOP. When the engine is first started and the RPM go above 400rpm (and is not in clear flood mode), the ECM enters OPEN LOOP fuel control operation. During open loop operation, the ECM ignores the O2 sensor input when making fuel delivery calculations and relies on the other sensors to determine fuel delivery to the engine.

There are specific tables stored in the PROM programming that contain the instructions the ECM uses to determine how much fuel to give the engine in this operating mode. Also during this period the ECM still constantly monitors the signal coming from the O2 sensor to see whether it is ready (hot enough) for closed loop mode.

A set of qualifiers are programmed into the PROM memory, which the ECM uses to determine when to enable closed-loop fuel operation. For example, qualifiers such as the coolant temperature and time from start, affect the O2 sensor output and the closed-loop fuel operation.

The amount of coolant temperature and elapsed time from startup qualifiers differ from engine to engine and differ a great deal, depending on whether the ECM is working with a heated or non-heated oxygen sensor.

For most non-heated O2 sensor applications, the general value for coolant temperature is 150°F and running time is 2.5 minutes. For systems using heated O2 sensors, these qualifiers are usually much less. When the system enters closed loop, the ECM still uses all other sensors/inputs to determine fuel delivery to the engine, but now it uses the O2 sensor inputs to adjust the fuel delivery based on what it sees in the exhaust.

The ECM looks at battery voltage and uses this information to compensate for variations in fuel pump output and injector response. This is needed because lower battery voltages cause the fuel pump to produce less fuel flow and causes the injectors to respond slower compared to what they would do at higher battery voltages. The ECM compensates for lower battery voltage by increasing the amount of injector on time (pulse width). This correction takes place in all operating modes.

The limp mode is in operation any time the ECM cannot operate normally. In this mode, the ECM looks to the chip to determine engine operation. The chip contains minimal information the ECM can use to allow the engine to run using only distributor reference pulses, throttle position, and coolant temperature inputs to change fuel and ignition timing calculations.

Limp mode was design to allow the vehicle to "limp-home" and not leave the customer stranded should a major problem occur. The ECM will implement back-up mode if anyone or a combination of the following conditions exist: the ECM input voltage goes below 9 volts. The Cranking voltage goes below 9 volts. The ECM internal circuitry fails to ensure proper Computer Operating Pulses.

Body control system - BCM

BCM computers are used to control various systems in the vehicle. In the past, a single computer was used to control the operation of all components functions. This BCM controls the operation of a multiple computers and modules in the vehicle.

The BCM controls system was designed to control the functions of automotive components such as service reminder and warning lights on the dashboard, interior and exterior lighting, security systems, locking or unlocking features, and other accessories.

Some examples of systems, which have not controlled directly by the BCM, are the ABS system, the engine management system and the transmission controller, or their individual modules. However, the BCM shares data with the computers in the vehicle and vice-verse.

The BCM monitor and control other processes, CPU outputs information and software executive decisions, so placed malfunction in one place can affect a number of different processes or systems. The BCM can interface with other components' operation as electric window or door monitor the results and take action based on this information. If the BCM does not function properly, it should be replace, not repaired.

The BCM has a network of sensors to transmit information about the operating conditions and durability of the components and accessories by feedback signals sent back to the BCM.

The Sensor signals are usually analog and a binary decoding are requires conversion to digital signals so the processor can understand and use it.

Analog converter-Digital produces binary digits to represent the sensor data. When the converter is ready with a binary number, it sends a signal to the processor that the conversion is complete and ready to send data.

Output commands are measure and sampled constantly as a method of sequentially monitoring outputs' commands to ensure they reach the right place. This usually achieved by a method called transitional sequential sampling.

Once a parameter is measured, the BCM begin processing it. It regularly performs tests to ensure information was sent to malfunctioning circuit. Input values are stored in a separate location.

Vehicle control computer will use all the components input as needed to make a decision. Following the data processing phase, the BCM send some output signals. For example, the computer can display the warning messages on the instrument panel and send information to the driver. The vehicle control computer also sends the information to the ECM to inform it of the operating conditions. Computers have several modes of operation: awake, asleep, and wake sleeping. The BCM receives regular reports on the mode of operation of other computer multiple times per second.

Logic modules

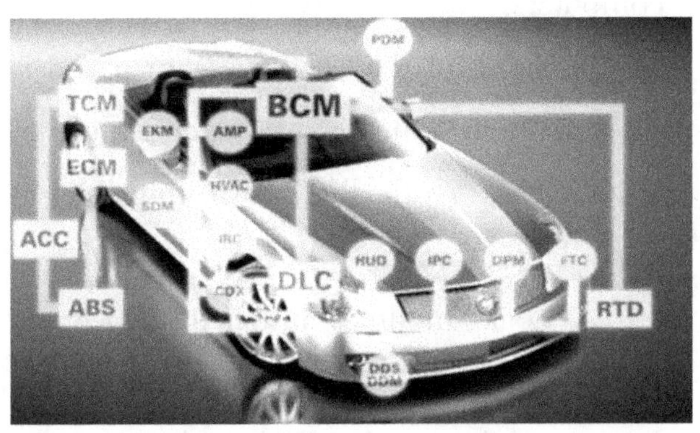

Logic modules has default program with predetermined strategy, so that each time there is a defect or missing information the default program is used. The Logic modules can send an output command, even there is limited or no data input in process. Logic modules have the ability to operate with failed components. Failures may include input/output information or communication errors.

Some Logic modules are

independent, meaning that they perform their duties without any contact with other computers.

Some of the characteristics of Logic modules that functions independently are only intricate with their own function, to control input/output.

Logic modules installed in a multi computer-control system. This means that Logic modules use data from various sensors, and switches, to control their connected components.

The control system of the vehicle can also control by lending outputs-information from the connected vehicle systems through a link-information-serial data link. The Logic modules evaluate information and control some of them by the output commands or cutting off connection.

Computers in multi- Logic modules vehicles can also inquire other Logic modules that are not directly connected to the by-wires and receives there status information. There specific task is to send messages through the serial data link.

Vehicle computers system is capable to perform functions and respond to the messages. The main strategy of vehicle multi-control system-is to avoid emptying the battery and maintain Good health. Hybrid electric vehicles equipped with multiple computers are connected in series; one computer is elected as a controlling computer or PMM. The PMM can be the vehicle BCM, or integration computer instrument panel. Check the manufacturer's instructions to determine the identity of the PMM for each vehicle.

Hybrid electric vehicles use an integrated power module on the dashboard called IPM, where the ignition start button is located, and a key operates the other module, so the PMM detects the vehicle status by using the state of the ignition signal circuits.

Most Hybrid electric vehicles have a limp mode activation command, or BPMM. Both PMM and BPMM receive command's activation on the serial data link. If the PMM, or the BPMM, do not accept the command, it enabled the default program. Computers react differently depending on whether they received data using serial data link or the ignition circuit.

During a limp-mode, computers receive data on the serial data link remain in the last command given to them and keep checking the serial data link for messages. If data indicate that the engine is running, the default value of computers running in ON. If the engine is not running, the computer goes into Closed-Off-awake mode, and the power system wills eventually go to shut down mode if there are no conditions that require it to be awake.

There are two exceptions to this rule are the engine control computer (ECM) and transmission of a TCM.

These computers will remain awake in their last action until and determine their status according to the status of the ignition system.

Computer software update

ECMs may need reprogramming for several reasons.

One is to fix factory bugs.

Reprogramming may be required if the factory settings for the OBD II self-diagnostics turns out to be overly sensitive - especially after a few years of operation. The same goes for drivability. Changing the fuel enrichment curve, spark timing or some emission control function slightly may be necessary to eliminate a hesitation, spark knock or other condition that develops over time.

When vehicle manufacturers calibrate the onboard diagnostics to meet federal emission standards, they have to draw the line somewhere as to what operating conditions might cause emissions to exceed federal limits 1.5 times.

That is where a fault code set and the Check Engine light comes on. It does not mean emissions really are over the limit, but it is possible based on laboratory dynamometer testing and field experience.

Depending on the application, the vehicle manufacturer may even set the limit a little lower just to be safe because the last thing any OEM wants is an expensive emissions recall.

Unfortunately, vehicle manufacturers do not always communicate their diagnostic strategies or even their operating strategies for their computerized engine control systems.

Some service manuals include a fair amount of system background information but others provide almost nothing beyond a basic diagnostic flow chart. The best advice when confronted with a troublesome code that keeps coming back or seems to set for no apparent reason is to check for any Technical Service Bulletins (TSBs) that may have been published. Chances are, it has been the programming of the ECM all along.

There are four steps computer for software updating process:

Locate the desired file calibration TIS (Technical Information System). The calibration files could be found in two areas: search by model and year technical service bulletins that address the problem in a particular product. You can search by reference to model under computer software update. Download the TIS calibration computer/scanner. Calibration updates Wizard (CUW) is an application for TIS calibration, responsible downloads files from TIS.

Make sure the battery is charging, to avoid any interference the update process.

You can download the files to your computer with your scanner. The computer software is reprogrammed. Make sure that the update is complete. Any modifications to the computer program have a registration sticker, usually under the hood.

Regenerative braking

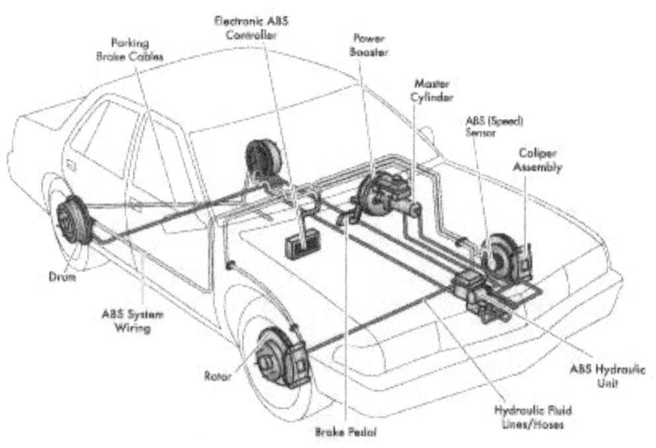

Hybrid electric vehicles are very different from traditional vehicles. The two main differences are the engine does not run at all times, and the high voltage batteries need to be recharged on the drive.

The control side of a hybrid brake system contains differences when compared to its conventional cousins to accommodate these conditions.

Energy cannot be created or destroyed. It can only be transformed from one form to another. Conventional friction brakes do this by transforming the energy of motion (kinetic energy) into heat energy, via friction, and then dissipating it into the air.

With regenerative braking, instead of just wasting that energy by releasing it into the air, the vehicle stores the energy and later converts it back into kinetic energy.

On a hybrid vehicle, the brake pedal

acts more like an interface for the ABS module than as a request for braking. The brake pedal and ABS module pressure sensors "tell" the modules of the hybrid system that a request for braking has been made and how much braking is desired – the driver's pedal efforts are calculated by the modules to produce the correct pedal feel and stopping power. The modules that control the hybrid system then apply regenerative braking via the traction motor inside the transmission as needed. More regenerative braking increases the amount of electrical charge generated for the hybrid battery while it produces more drag to slow down the vehicle. Of course, at a stop or near stop, the friction is applied to stop the vehicle from rolling. Some hybrid systems apply the hydraulic brakes at lower speeds.

It is normal for a hybrid to wear the rear brake pads two and three times faster than the front pads because, during many light-braking events, the rear friction brakes will be lightly applied and the front may not be applied at all until the vehicle is at a stop or near stop.

On most hybrids, the pressure generated for the calipers is not the direct result of the driver's foot pedal efforts. The pedal-generated pressure is used to stimulate a pressure sensor in the system that makes up the brake request. Normally, the fluid pressure for the wheel hydraulics is generated by the pump and/or accumulator in the ABS hydraulic control unit or a stroke simulator and is regulated by the teamed efforts of the ABS and the ECM's logic.

The higher the request for braking, the system electronically calculates and decides how to apply friction and regenerative braking to meet the demand. All brake applications, from feather-light braking to a panic stop, are electronically calculated and delivered from a normally operating electrohydraulic brake system.

In the event that the ABS hydraulic control unit cannot function well enough to build pressure, the driver's pedal pressure will directly provide the hydraulic pressure for the calipers. There are valves inside the hydraulic control unit that allow direct passage from the pedal feel simulator, through the hydraulic control unit and to the wheels. This is called "manual mode." In manual mode, there will be no boost so the driver will have a very hard/stiff brake pedal. The driver will need so much braking effort in this situation that it may prompt them to tow the vehicle for fear of having no ability to stop at all. Complaints of "no brakes" or "barely stops" is the often how customers will describe a brake system that has entered manual mode.

* * * * *

Replacement of friction material is the same for a hybrid as it is with a conventional system, however, the ABS module will test the brake hydraulic system's integrity by pressurizing it during key-off events (door opening, dome light activation, etc.) and again when the key is switched on.

After four minutes have passed since the key was switched off, the accumulator discharges its stored fluid pressure back into the master cylinder reservoir. Before attempting friction material replacement, you must follow the procedures to depressurize the system.

It is common to come across a hybrid with 70,000 or 100,000 miles with the original set of front

Always use a high-quality pad for

hybrid applications to avoid problems. It is not a question of better performance, but more a question of quality and engineering.

Since it is impossible to perform a conventional break-in/bedding procedure on the test drive, make sure the manufacturer promises excellent performance right out of the box. In addition, applying a non-direction finish with a ball hone will help the new pads evenly deposit a layer of friction material to the new rotor.

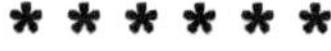

The two-mode hybrid system has a unique brake pedal because the hydraulic system is not the lone provider of vehicle deceleration. How the system blends these forces can cause a customer to notice the changes and become concerned.

Over several braking events, the customer may perceive that he is applying a consistent braking force and the vehicle may decelerate at slightly different rates. This is a normal operating characteristic. The driver may think it is the same type of stop, but to the system, a little extra speed or pressure can change braking performance or the blending of the brake mechanisms.

Another common complaint for used hybrids is noise and pedal feel. Most two-mode Hybrid electric vehicles use a braking system that employs a high-speed electric motor/pump. Pump operation sometimes can be felt in the brake pedal as the system builds pressure. This is most noticeable to customers who start the vehicle with their foot on the brake pedal. If the engine not running the noise become noticeable as the driver use the brake pedal.

Whenever the electric motor rotates faster than the computer affecting him, it does not function as a motor but a generator.

Instead of taking power from the inverter and converts it into mechanical energy, the electric motor converts a mechanical energy and returns it to the inverter. The inverter transfers this energy and charge the battery-HV high-voltage.

This process usually occurs when the force of gravity contributes to the movement of the vehicle on a slope.

When energy is restored back to the inverter, is absorbed first into the super capacitors causing increasing DC voltage.

The inverter monitors the voltage equity, and when the voltage rises above the level recommended the inverter to protect the internal components. When that happens, no motor and charging system is under control, they begin to slow down to a stop. To optimize the performance and reliability of the braking energy recycling, manufacturers use dynamic braking techniques to manage the energy transition through the inverter circuits, for optimal battery charging.

The most common method of dealing with excess braking energy is dynamic. This is a more efficient solution and offers higher performance. Dynamic brake controller moves the excess energy into the counter to convert the heat energy when the voltage rises, to protect the internal components. Electronic brake controller can quickly change the frequency and thus control the power and speed of electric motors. To optimize the performance and reliability of the electric motor, braking stromal used.

Breaded braking is software that can change the electrical voltage electric motor to cause a temporary stop to be very inefficient.

The recovered energy is lost in the electric motor itself as heat. This is an effective way to deal with the low amount of energy is restored, if the duty cycle duty cycles are high, instead of using 100% friction brakes. Since the braking energy recycling occurs when the electric motor runs at a higher speed than in a control computer, there are several ways to deal with the negative consequences of excessive energy.

One way is to eliminate it, by increasing the speed commands from the computer so that it is equal to or greater than the speed of the actual engine. However, most of the time the computer may change the voltage of the motor, but it is usually not practical because of the current inertia will not prevent the actual speed change. This option is good for situations where the recovered energy is low, but high frequency. In some cases there to stop the energy and let it be wasted as heat, is divided between the two electric motors. The amount of the recovered energy depends on various factors, including vehicle weight, speed, and the rate of slowdown. Hybrid electric vehicles spread the use of energy-reconstruction-user serial braking system electro-hydraulic braking. Electromechanical braking system includes a hydraulic control unit, which controls the hydraulic pressure of the brake cylinders, as well as balancing the brake operation. In Toyota PRIUS first generation, there are four pressure sensors of the braking system and two pressure switches. The system must be designed to allow proper use of the ABS antilock braking system. Electronic brake controller determines the amount of braking energy recycling in relation to the amount of friction braking. To maintain a sense of the brake pedal of hybrid electric vehicles as ordinary vehicles, Toyota and Honda hybrid, the braking energy-recycling mode and brake operation is controlled by the brake pedal.

In the first part of the course of the brake pedal powered energy recycling only, and then when the pressure on the brake pedal mounting, brake pads operate. This causes the brakes to reduce erosion and hydraulic system. When the brake pedal is pressed, the amount of energy recycling computer is determined by the difference between the initial value and the value of "load on the brake pedal."

Honda CIVIC has an integrated control of the brake pedal without the two modes. If the driver steps on the brake pedal gently, moving the cylinder plunger will be small hydraulic brake pressure is also small and energy recycling operation is applied gently.

As the cylinder pressure increases, so is the recycling of energy, which applied automatically.

Hybrid vehicle braking energy recycling takes place not only when the brake pedal is operated. When the accelerator was pressed full-gently released purpose of slowing down, recycling-energy-containment will occur automatically. Toyota PRIUS equipped with an LCD screen that shows how much -watt-hour accumulated every 5 minutes. Illustration of a small sun appears on the display screen, and a sun-lit figure is 50-Watt/hours. When one sun appears, there is enough battery power to operate the bulb 50 W for an hour

Mode B Gear

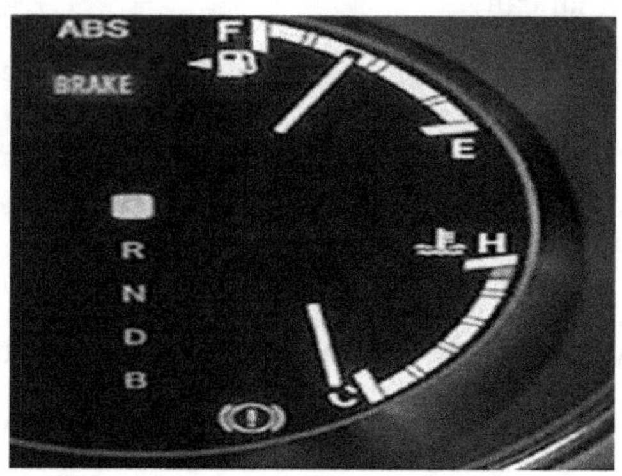

Toyota Hybrid electric vehicles have a gear condition marked "B". This situation increases the engine braking (higher engine revolutions), when the vehicle is descending in a long, steep slope to prevent the battery from be overcharged in higher rate.

Stop and Go operation

At a full stop, like at a red traffic light or stop sign, the gas engine usually shuts off to eliminate idling and reduce emissions. The electric motor is now ready to propel the car when push on the gas pedal. In crowded cities with lots of stops and go traffic, this can save you a lot of fuel.

The electric motor assists in acceleration from stop by using power from the high voltage battery. With hybrids vehicles, in a stop and go traffic, the most fuel is saved, unlike regular cars where you burn the most fuel. The gas engine turns on and off as needed while you drive. Rapid accelerations will still hurt gas mileage, just like a regular car.

When a hybrid vehicle speed drops below 30 MPH, the Stop and Go the ICE generally remains off until the car accelerates to start again from a Stop.

Stop and Go situation will not occur in the following conditions:

- The ICE is not yet warmed up.
- The transmission is in reverse.
- The HV battery is not charged enough.
- The weather condition keep the ICE cold in Stop and Go traffic.

In addition, the air conditioner can keep the Stop and Go from taking place when the 'Auto' mode is selected for the air condition; the ICE may continue to run to support the compressor operation. If the "Econ" is selected, the Stop and Go may take place, causing the air condition compressor to stop temporarily.

The Air condition

Standard compressor driven by a belt only and have only two modes: either it runs at full power, or it turned off completely. The compressor runs at 100% output until it reaches, then turn off the desired temperature. When the temperature varies the compressor starts again. Such activity creates waste of energy, increased wear, and relatively high power consumption. The innovative compressor in some Hybrid electric vehicles has a variable output and the ability to operate at all ranges between off or 100%. With the operation of the air condition, the compressor starts at full power until the desired temperature achieved. When this is achieved, the electronic system reduces the compressor output to maintain the desired temperature. because the system does not have an ignition key, you can save close to 30% of the energy.

To allow function of Stop and Go in hybrid electric vehicle, when the air conditioning is on, the air conditioner compressor powered by a high voltage with or without conventional drive belt. Those Electric compressors can be powered electrically only.

Drivetrain

Variable transmission - CVT

Ordinary automatic transmission has a certain number of gears. Each gear is good for limited range, so when you keep accelerating, the transmission shift up through the gears, first 1, then 2 and so forth. A continuously variable transmission (CVT) technically does not have gears at all. The CVT is a gearless transmission, which can change seamlessly through an infinite number of effective gear ratios between maximum and minimum values. This contrasts with other mechanical transmissions that offer a fixed number of gear ratios. The flexibility of a CVT allows the input shaft to maintain a constant angular velocity.

A belt-driven design offers

approximately 88% efficiency, which can be offset by lower production cost and by enabling the engine to run at its most efficient RPM for a range of vehicle speeds. The ratio of the CVT can be changed to allow the engine to produces greatest power. a CVT does not strictly require the presence of a clutch.

Nissan has used CVTs across its lineup for years, and Honda and Subaru have followed suit.

There are two main designs of sequential transmission. Split system-power used by Toyota/Lexus and Ford Motor System-belt and pulley, used by Honda.

Variable-diameter pulley

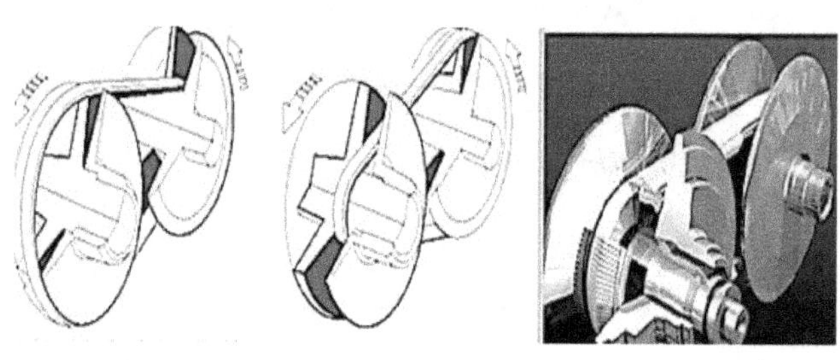

In this most common CVT system, two V-belt pulleys split perpendicular to their axes of rotation, and a V-belt running between them. The gear ratio is changed by moving the two sheaves of one pulley closer together and the two sheaves of the other pulley farther apart. Due to the V-shaped cross section of the belt, this causes the belt to ride higher on one pulley and lower on the other. Doing this change the effective diameters of the pulleys, which in turn changes the overall gear ratio. The distance between the pulleys does not change, and neither does the length of the belt, so changing the gear ratio means both pulleys must be adjusted (one bigger, the other smaller) simultaneously to maintain the proper amount of tension on the belt. The V-belt needs to be very stiff in the pulley's axial direction to make only short radial movements while sliding in and out of the pulleys.

Steel reinforced v-belts are sufficient for low-mass low-torque applications like utility vehicles and snowmobiles but higher mass and torque applications such as automobiles require a chain. Each element of the chain must have conical sides that fit the pulley when the belt is running on the outermost radius. As the chain moves into the pulleys the contact area gets smaller. As the contact area is proportional to the number of elements, chain belts require many small elements. The pulley-radial thickness of the belt is a compromise between maximum gear ratio and torque. For the same reason the axis between the pulleys is as thin as possible. In a chain-based CVT, a film of lubricant is applied to the pulleys. It needs to be thick enough so that the pulley and the chain never touch and it must be thin in order not to waste power when each element dives into the lubrication film. Additionally, the chain elements stabilize about 12 steel bands. Each band is thin enough so that it bends easily. If bending, it has a perfect conical surface on its side. In the stack of bands, each band corresponds to a slightly different gear ratio, and thus they slide over each other and need oil between them. Also the outer bands slide through the stabilizing chain while the center band can be used as the chain linkage.

A subset of CVT designs called infinitely variable transmissions IVT, in which the range of ratios of output shaft speed to input shaft speed includes a zero ratio that can be continuously approached from a defined "higher" ratio. A zero output speed (low gear) with a finite input speed implies an infinite input-to-output speed ratio, which can be continuously approached from a given finite input value with an IVT. Low gears are a reference to low ratios of output speed to input speed. This low ratio is taken to the extreme with IVTs, resulting in a "neutral", or non-driving "low" gear limit, in which the output speed is zero, unlike the neutral position in standard transmission,

There are several locations and several measures, which, the computer detected failure in the transmission, and inform the driver that service is needed. On Honda hybrids, the D display will start blinking on the dashboard.

In addition, the transmission will default to Limp- Home mode to allow the driver to reach a nearby service dealer.

Transmission control

Although electronically controlled transmissions function in the same basic way as the hydraulic ones did, the computer adds a new level of efficiency.

Most modern electronically controlled transmissions do not have governors anymore or even throttle pressure valve devices. The decision on what gear the transmission to select is now executed by the transmission control module that activates shift solenoids on the valve body.

At the heart of any electronic drive train control would be the transmission control module. The TCM relies on programming stored in its memory to provide gear shifting at the optimal time.

The first input the controller looks at to determine what action to apply is the position of the shifter. When the drive range is selected, the control module will use its shift schedule to decide when to up shift automatically. The computer also looks at the engine temperature and the load of the engine as determined by the map sensor, the speed of the vehicle determined by the vehicle speed sensor and attention to the throttle position selected by the driver is obtained through the TPS sensor.

The transmission controller also monitors a few other switches and sensors. An important one is the brake switch, which is used to disengage the torque converter clutch when the brakes are applied.

the signal from the brake switch is sent to the control module that down shifting for engine braking is needed.

One of the slicker electronic transmission controls is the ability of the TCM to learn the habits of the driver and provide a shift schedule that suits the driver's needs. The transmission control module does this by monitoring and remembering the typical driving style of a driver and the operating conditions of the vehicle at the time.

With this information the TCM can adjust the timing of the shifts and the operation of the torque converter clutch to provide good power and smooth shifting at the right times. When electronic systems are capable of learning and storing memory in this manner, they will be considered adaptive or smart.

The TCM is also able to compensate for internal problems such as seal leakage and wear and tear on clutch discs and friction plates.

When the transmission can no longer compensate for an internal problem, it will set check engine light codes in the P0700 range.

Transmission oil pump

The pump draws fluid through the pipe and filter basin transmission. Since the torque converter rotates with a motor fuel, hydraulic pressure pump is available only when the gas engine is running.

To adjust the transmission in hybrid electric oil pump is used to maintain oil pressure-Transmission in Stop and Go when the engine Stops its operation.

After starting a new engine, an electric pump is Stopped and the transmission oil pressure again supplied by the oil pump hydraulic transmission.

The transmission's oil pump is powered by a three-phase alternating current electric motor, controlled by a control module. The oil pump is mounted outside the transmission. There are three wires connecting the oil pump motor to the control module.

Toyota Power Split System

The heart of the Toyota Prius Synergy Drive is a simple little device called the Power Split Device. A planetary gear set removes the need for a traditional transmission components and acts as a continuously variable transmission (CVT) but with a fixed gear ratio.

The Power Split Device allows the car to use power from an internal combustion engine (ICE), as well as the two electric Motor/Generators (MG1 and MG2), all spinning at different and variable speeds. The Power Split Device allows the smaller of the two Motor, the MG1, to act as a starter for the ICE, thereby eliminating another component of a traditional engine.

MG2 is the larger of the two electric motor/generators and is referred to as the traction motor because its speed (RPM, or revolutions per minute), has a fixed relationship to the speed of the wheels. Rotation speeds of MG1, MG2, and ICE are inter-dependent, and the speed of MG1 will always change when you vary the speed of either of the other two. MG1 has a maximum speed of 10,000rpm in either direction (positive or negative) with a 6500-RPM low limit if the ICE is off. This lower limit means the ICE will always spin if the vehicle is driven above 42mph. the MG1 often does change its spin directions under normal driving conditions.

The ICE rotation is limited to speeds between 1000rpm and 4500rpm. It can also stop completely, but anything between 0 and 1000 will force the slider up or down. That is because the engine would not be able to operate effectively below that speed. The hybrid computer will stop the ICE when it decides and start it again when more power is needed, or a higher speed from MG1.

The basic method to split the drive is very simple and is built around two motors/generators, electric and the engine. When the vehicle stops, the engine and the two electric motors also cease.

When the vehicle is running at low speed with light acceleration, the MG2 motor is the only one to drive it. When the vehicle is moving at high speeds, the engine must be running and its power is integrated with the one of the motor MG2.

Power split device

The power split device is a planetary gear set. The electric motor is connected to the ring gear of the gear set. It is also connected directly to the differential, which drives the wheels. Therefore whatever speed the electric motor and ring gear spin at determines the speed of the car.

The MG1 motor/generator is connected to the sun gear of the gear set, and the ICE is connected to the planet carrier. The ring gear's speed is totally depended on the other three components of the motor, and together they control the output speed.

When you accelerate, initially the electric motor and batteries provide all the power. The ring gear of the power split device is connected to the electric motor, so it starts to spin with the motor. The planet carrier, which is connected to the engine, is stationary because the engine is not running. Since the ring gear is spinning, the planets have to spin, which causes the sun gear and generator to spin. As the car accelerates, the generator spins at whatever speed it needs to in order for the engine to remain off position. Planetary System Device is used as a power splitter.

MG1 and MG2 motors function as

electric motors and generators. MG1 and MG2 are the additional power source that provides power assistance as required.

MG1 is used as a starter for the ICE. The MG1 charge the HV battery and provides electricity to MG2 motor. MG2 motor used as a driving power that and help the ICE.

Driving the hybrid vehicle

Starting off

Current HV battery passes high voltage through the converter and run the MG1 motor clockwise to start the Internal combustion engine at up to 1000 RPM.

After the Internal combustion engine begins to run, the MG1 motor reverses its direction of rotation and acts as a generator. At this point the driving force is being 'split' between the ICE and MG1.

The power generated by the MG1 directed at the MG2 to assist drive the vehicle, or used charge the HV battery - as necessary.

Crawling

A conventional car with automatic transmission will tend to move forward from a stop if you take your foot off the brake. This is probably a side effect of the torque converter operation, but has the benefit of preventing the car from rolling backwards on hills while you transfer your foot to the accelerator. It is called "crawling". Toyota wants the car to feel familiar to drivers. Therefore, creeping is simulated too. A small amount of power from the battery is applied to MG2 when the driver releases the brake. This moves the car gently forward.

If the driver applies a bit of accelerator pressure, the power to MG2 is increased and the car will move forward more positively. Since MG2 is quite powerful and has high torque, the driver can take-off on electric power only up to a fair speed as long as the traffic allows him to accelerate gently. The more pressure on the accelerator, the sooner the ICE will start to help with its torque and the electricity generated by MG1. If the driver pound the pedal hard to the floor, the ICE will fire up right away, although it will be off before it run- up to speed, to contribute more power. For most around-town starts, the vehicle using just the drive power from MG2. The ICE remains stopped and MG1 spins freely backwards.

Slow Driving

If you reach your target speed before the ICE fires up, you can continue to drive using electric power only. This is called "EV mode", since the car is powered in exactly the same way as a pure electric vehicle (EV). The ring gear turning as the MG2 powers the car, the planet carrier and ICE stopped, and the sun gear and the MG1 are spinning freely backwards.

Even if the ICE does start during acceleration, when you get up to speed and reduce accelerator pedal pressure, the power needed may drop to a level that MG2 can comfortably provide. The ICE will then turn off and EV mode will start again.

Cruising at Moderate Speed

Once the vehicle reached a steady speed the ICE power supply get drops and runs at a low spin rate. The MG1 must now spin backwards. By spinning backwards, it makes the planet gears rotate forwards. The rotation of the planet gears are added to the rotation of the planet carrier, making the ring gear move faster.

MG1 must exert a backward torque at the sun gear. This is how the ICE achieves the power to turn the ring gear forward. Without the resistance of MG1, the ICE would just spin up MG1 instead of moving the car. Therefore, the inverter electronics must draw power from MG1 to charge the high voltage battery.

This causes the planets to spin forwards and add more spin to the ring gear. The ring gear still gets only 72% of the torque of the ICE, but the speed at which the ring spins is increased by MG1 motoring backwards. Turning the ring faster allows the car to travel faster for a low ICE RPM. The MG2 resists the car's motion slightly with its generator drag and produces electricity to be fed to MG1. The ICE remaining mechanical torque drives the car.

The Energy Monitor display will show engine power to the wheels and the motor charging the battery. It may also alternate between charging and drawing from the battery through the motor to power the wheels.

Cruising

Anytime the driver let go of the accelerator, the vehicle considered cruising. The engine does not try to push the car forward. The car slows down gradually due to rolling resistance and aerodynamic drag. In a conventional car, the engine is still connected to the wheels by the transmission. The engine turns over without fuel and therefore tends to slow the car. This is so called engine braking. Although there is no reason for this to happen in the Prius, Toyota has decided to give the car the same feel as a conventional car by simulating engine braking. When you coast, the car slows faster than would be the case if only rolling resistance and aerodynamic drag were acting on it. To produce this additional slowing force, the

MG2 is acting as a generator and charges the battery. Its drag simulates engine braking. Because the ICE is not needed to power the car, it stops. The MG2 motor acts directly at the ring gear, the planets spin forwards and MG1 spins backwards. The MG1 motor does not produce power and does not generate power. It just spins free.

Maximum acceleration

When the power demand is high, the ICE and MG2 both contribute torque to drive the car in much the same way as described above for starting. As the speed of the car increases, the torque that MG2 is able to supply is reduced because it begins to operate at its 33kW power limit. As a conventional car speeds up, the step gearbox up shifts and the torque at the axle is reduced so that the engine can drop its rotation rate to a safe RPM value. Although it does so using a completely different mechanism, the Prius gives the same general feel as a conventional car accelerating. The main difference is the complete absence of lurching as the step gearbox shifts because there simply is no step gearbox. (Cars with true CVTs, such as the CVT Insight, also have a smooth feel during acceleration.)

Seventy-two present of its torque goes mechanically through the ring gear to the final drive and the wheels. twenty eight percent of its torque goes to MG1 via the sun gear where it is turned into electricity. This electricity powers MG2, which adds some extra torque at the ring gear.

The harder the acceleration, the more torque the ICE will produces. This increases both the mechanical torque though the ring and the amount of electrical power generated by MG1 for MG2 to add more torque. The inverter might draw extra power from the battery to boost MG2's contribution. If the power demand is not that high, some of the electricity produced by MG1 may be used to charge the battery, even while accelerating. When there is a greater need for acceleration, the ICE and MG2 motor provide propulsion.

During deceleration or by using the brakes in the vehicle, a situation in which braking energy regenerative engine, the MG2 is running as a generator, the ICE is off and MG1 and MG2 passing the current charge through the converter to charge the high voltage battery.

The MG2 is the main driving force. When the vehicle is running on battery power only, MG2 is one driving it.

Downhill Driving

When the driver decelerates gently or goes downhill, the power needed to keep the car moving is reduced, because of gravity and inertia. Therefore, he decreases the accelerator pedal pressure. Depending on speed, the ICE may stop delivering any power at all if MG2 can supply what is needed. In Slow Driving, the MG2 could supply all the needed power with the ICE stopped. When accelerating and driving at constant speed on the flat road, this operation is unlikely to occur at speeds above 40 M.P.H. because the power needed to overcome aerodynamic drag is enough to cause the ICE to start. Under some conditions, the EV mode can occur at a higher speed, and is quite likely to occur when slowing down or driving fast downhill.

To operate in EV mode at speeds above 42 M.P.H., the car has to protect MG1 from over spinning in the same way as for Coasting, above. The only difference is that instead of the car's motion turning the Power Splitting ring gear, it and the car are powered by MG2. MG1 still generates power to resist excessive spin with the result that the ICE turns over. No fuel is supplied and no sparks are generated. Of course, by doing this, MG1 is drawing off power that otherwise would drive the car. Some is lost in spinning the ICE, but some shows up as electrical power generated by MG1.

Braking

When the driver want to slow the vehicle faster and press on the brake pedal in a conventional car, this pressure is transmitted by a hydraulic circuit to the brake pads at the wheels. The brake pads rub against metal disks or drums and the energy of motion of the car is converted to heat as the car slows down. The Prius has this exact same braking mechanism, but it has something else as well - regenerative braking. Whereas during coasting MG2 produces some generator drag to simulate engine braking, when the brake pedal is pressed, the electrical power generation of MG2 is stepped up and generate much greater drag to contribute to slowing the car. Unlike friction brakes, which waste the car's kinetic energy as heat, the electrical power produced by regenerative braking is stored in the battery and will be re-used later. If the driver moves the mode selector lever to the "B" position, the engine will be used for engine braking. Whereas normally the engine is stopped during braking, in this mode the computer and motor/generators arrange for it to turn over without fuel and with an almost closed throttle. The resistance it offers slows the car, reducing brake heating and allowing you to ease up a little on the pedal.

Reversing

The Prius has no reverse gear that would allow the ICE to push the car backwards. Therefore, it can only move backwards under electric power from MG2. No direct help from the ICE is possible. In most cases, the car will stop the ICE when you put the selector lever in the R position. As MG2 turns the input to the reduction gears backwards, the Power Splitting ring gear will also move backwards. With the ICE, and therefore the planet carrier, stopped, this simply means that MG1 will turn forwards. It turns free, without using or generating power. This is exactly like EV mode, but backwards. If the battery state of charge is low when the selector lever is in the R position, then MG2 still simply drives the car backwards as before. The only difference is that with the planet carrier running forward, the sun gear and the MG1 will spin faster forward and the computer must limit the backward speed of the car to a lower value to protect MG1 from over-spinning.

Idling

When the car at rest the entire power train the gas/petrol engine, the electric motors and the generator are automatically shuts down. No energy is wasted by idling.
If the battery charge level is low, the gas/petrol engine is kept running to recharge it. In some cases, the gas/petrol engine may be turned on in conjunction with the air-conditioner switch operation.

Vehicle operations phases:

Fhase1: ICE is cold. You cannot start with electric motor alone. The engine needs to run and warm up.

Phase 2: When the Internal combustion engine system reaches 40 degrees, the computer switches to Phase 2 and you can drive with electric motor only.

Phase 3: when the engine reaches a temperature of 73F degrees, the computer goes to phase 3. The ICE will not stop, unless the vehicle comes to a full Stop for 5 or 10 seconds. When you reach to 32MPH, and begin to slow down while pressing the brake pedal, the computer returns to phase 2, the engine will stop immediately and the vehicle will operate on electric motor only.

Phase 4: is usually normal, full hybrid operation. Above 35 MPH, the ICE must help and provides propulsion power. The maximum speed of the electric motor operation mode is 32 - 38 MPH.

By-wire Technology

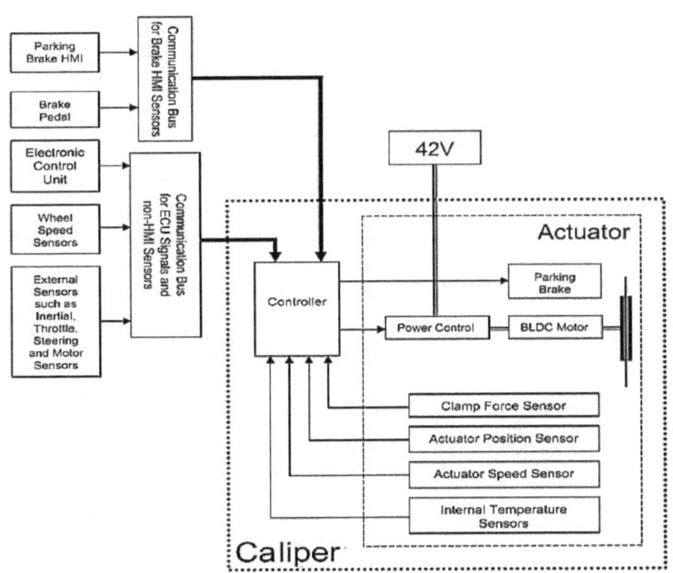

Drive by wire, Steer-by-wire, or X-by-wire technology in the automotive industry is the use of electrical or electro-mechanical systems for performing vehicle functions traditionally achieved by mechanical linkages. This technology replaces the traditional mechanical control systems with electronic control systems using electromechanical actuators and human-machine interfaces such as pedal and steering feel emulators. Components such as the steering column, intermediate shafts, pumps, hoses, belts, coolers and vacuum servos and master cylinders are eliminated from the vehicle. This is similar to the fly-by-wire systems used widely in the aviation industry.

Examples include electronic throttle control and brake-by-wire.

Response times improved for drive by wire systems through elimination of mechanical linkages.

Electronic fuel injection metering in gasoline engines is now widely used. Electronic throttle control is also widespread. Purely electronic brake and steering systems will find widespread application in passenger cars.

Hybrid electric vehicles employ limited electronically controlled regenerative braking, but the standard hydraulic braking system is retained.

Electronic throttle control system helps accomplish vehicle-driving force by means of an electronic throttle without any cables from the accelerator pedal to the throttle valve of the engine. In electric vehicles, this system controls the electric motors by sensing the accelerator pedal input and sending commands to the power inverter modules.

A pure brake by wire system eliminates the need for hydraulics completely by using motors to actuate calipers and lock the wheels in comparison to the currently existing technology where the system is designed to provide braking effort by building hydraulic pressure in the brake lines.

The direction of motion of the vehicle (Forward, Reverse) is set by commanding the actuators inside the transmission through electronic commands based on the current input from the driver (Park, Reverse, Neutral or Drive).

The Electric Power Steering controls a car with less mechanical components/linkages between the steering wheel and the wheels. The control of the wheels' direction is established through electric motor, which is actuated by electronic control units monitoring the steering wheel inputs from the driver.

Electric Power Steering can be considered as a stage of evolution from mechanical steering to steer by wire systems.

The parking pawl in a traditional transmission has a mechanical link to the gear lever and locks the transmission in the park position when the vehicle is set in Park. It is different from the parking brake. A park by wire system uses electronic commands to actuate the parking pawl by a motor when the driver puts the vehicle in park. The benefits of electronic technology application are improved performance, safety and reliability.

Throttle Position Sensor

TPS

A throttle position sensor (TPS) is a sensor used to monitor the throttle position of a car. The sensor is usually located on the throttle and monitors its position.

The accelerator pedal often includes two position sensors. The accelerator pedal sensors are used in electronic throttle control or drive by-wire systems and use those sensors for kick-down function on automatic transmissions.

Modern day sensors are non-contact type. These modern non-contact TPS include Hall Effect sensors, Inductive sensors, magneto resistive and others.

In case of failure of the TPS operation, the Check Engine light remains illuminated even if there is no problem or error in the ECU. It cannot be corrected by clearing ECU errors by running diagnostic software to rectify the malfunction the TPS needs to be replaced.

When the driver presses, the accelerator pedal is ' feel ' resistance designed to provide the same feeling as the conventional acceleration pedal.

By-wire Steering System

Hydraulic steering systems have long dominated the vehicle market because of their familiarity both to vehicle designers and to operators. More recently, the trend is towards the use of electronic steer-by-wire systems that provide greater design flexibility by enabling software to customize the connection between the steering wheel and steering mechanism. The latest generation of integrated steer-by-wire systems consumes less power, is less expensive, and offer the ability to be programmed to provide a wide range of value-added features, Moving away from hydraulic steering Hydraulic system. Another reason is substantial performance improvements that were made in electric motors in recently years.

Electronic steering systems provide nearly maintenance free operation and are thus much less prone to fail due to lack of maintenance.

Electric power steering in Hybrid electric vehicles is in use to eliminate the need for hydraulic pump. Electric steering are more effective than conventional power steering, since the electric motor assist only when the wheel is turned on, while the hydraulic pump must run all the time.

The level of assistance can be easily planned by vehicle type, road speed, and driver preference.

Electric power Steering

Electrically assisted power steering (EPS) is the latest technology, replacing hydraulic assist system. Someday every car control will be by-wire; today's EPS looks like a step in that direction.

The power steering system is operates by an electric motor and the operation of the steering wheel is still active even when the engine is off. The computer uses the data from the EPS sensor, Yaw sensor and the skid control system to determine and turn the electric power steering motor accordingly.

Honda electric power steering

If the computer detects EPS system failure it sends a warning to the driver by activating a warning light. The computer saves the fault code and the system power turns off, however, the system still provides the option to drive manually.

Electrically assisted power steering from Honda hybrid cars uses an electric motor to provide steering assistance and replaces the need for a hydraulic pump, and hoses.

The Toyota Highlander and Lexus RX 400h use electric power steering is different way due to differences in size of the Hybrid electric vehicles. This Electrical assisted power steering unit uses a brushless motor direct current motor assembly located on the steering rack. Electric power steering (EPS) provides auxiliary power even when the engine is off. It also improves fuel consumption that is light in weight and engine consumes energy only when required for steering assistance force.

The steering system consists of the following components: The electric DC motor, a reduction relay and torque sensor.

Power steering Torque sensor

The power-steering control unit uses a torque sensor as the main input for determining the amount of steering assistance needed. The steering column is split into two parts: The input shaft, from the steering wheel to the torque sensor, and the output shaft from the torque sensor to the steering shaft coupler. The input and output shafts are separated by torsion bar, where the torque sensor is located. The torque sensor measures the shift angle between input and output shaft.

The sensor by itself is split into two parts, the rotor and the stator. The rotor, a multiple magnet ring, is mounted on one side of the shaft.

The Hall sensor is positioned over the stator. The sensor detects when torsion occurs between input and output shafts, which is proportionate to the real shift angle.

This information is processed by the EPS ECM, which controls the power steering 3-Phase motor.

The Torque Sensor was developed for vehicles with electric power steering EPS. The sensor measures the steering force applied by the driver and thus enables sensitive control of the electric steering support. The sensor is based on a contactless magnetic measuring principle. It consists of the Magnet Unit, the Flux-Tube Unit and the Sensor Unit.

The Power steering Torque sensor and the angle sensor integrate into the electric power steering and provide the driver with two feedbacks simultaneously: records the torsion bar angle required for the steering movement and measures the angle and speed of change in the position of the steering wheel.

Both sensor functions are integrated within one sensor.

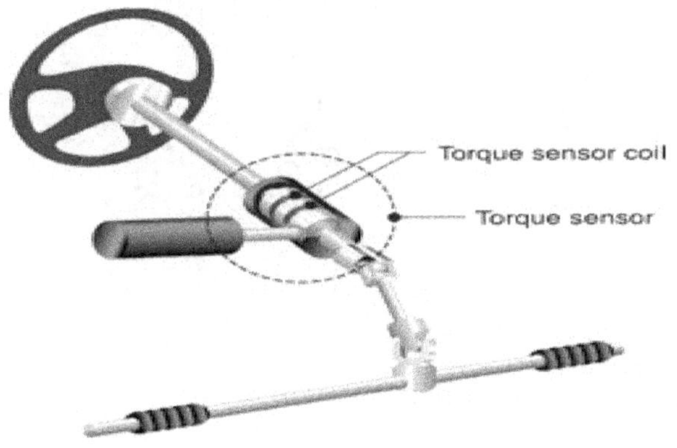

The power steering torque sensor is non-contact that detects the movement and the torque exerted on the wheel torsion rod.

Calibration of torque sensor is required at any time once one of the following assemblies removed or replaced the wheel orientation system. (Containing the torque sensor), the ECM computer, the steering wheel, and if there is a difference between left and right effort to steer the wheels.

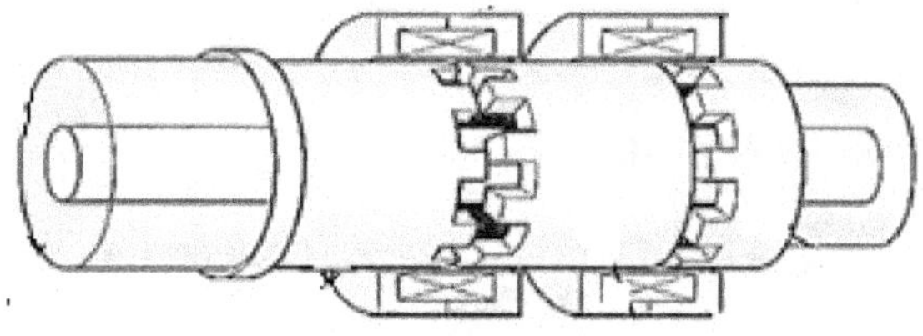

By-wire Brake system

There are actually two types of brake-by-wire systems. Hydraulic brake-by-wire system uses additional hydraulic parts to pressurize the brake disks. Electric, or "dry," brake-by-wire, on the other hand, simply uses an electric motor and no hydraulic brake fluid.

Computers and small electric motors control the brake systems, which eliminate the need for brake fluid, as well as most of the mechanical parts associated with conventional brake system.

Electromechanical brakes are considered to be easy task for production, and maintenance, and environmentally friendly.

The ABS system electronically

monitors the rotation of the wheels, and when it detects a wheel lock, it immediately releases the brakes for a split second so that wheel returns immediately to action.

Braking by-wire technology in hybrid electric vehicle is becoming more widespread because the conventional hydraulic braking systems does not offer an easy way for the regenerative system to function well, which by itself is a challenges when trying to increase the amount of kinetic energy.

The Braking by-wire system can respond in hundredths of a second and great accuracy by controlling the brake pressure on each wheel.

Even the ABS and ESP function more efficiently since they are integrated with the brakes by-wire system.

Hybrid

Brake system

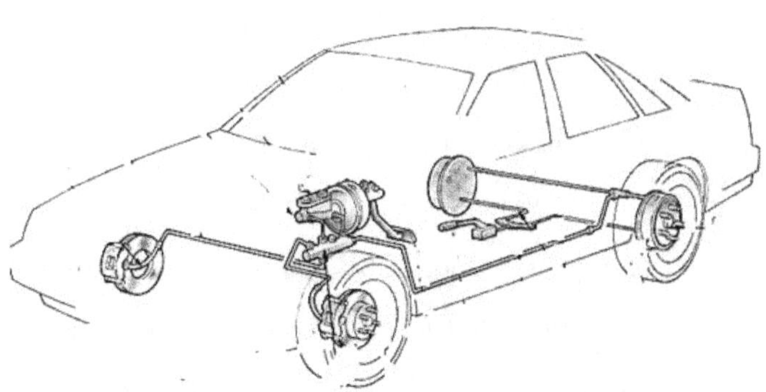

Electro-Hydraulic Brake

In a hydraulic brake system, the driver applies force by a mechanical link from the pedal to the master brake cylinder. In turn, the master brake cylinder develops hydraulic pressure in the wheels. In contrast, the electro-hydraulic brake provides the brakes with a brake fluid supply from the hydraulic high-pressure reservoir, which is sufficient for several braking events. A piston pump driven by an electric motor supplies a controlled brake fluid pressure between 140 and 160 Bar in the gas diaphragm reservoir.

When the driver presses the brake pedal - or when ESP intervenes to stabilize the vehicle - the he electro-hydraulic brake control unit calculates the desired target brake pressures on each individual wheel. With independent pressure modulators, the system regulates the hydraulic pressure at each wheel. These four pressure modulators consist of one inlet and one outlet valve, controlled by electronic output stages.

The system employs a travel sensor and a pressure sensor at the pedal to measure the speed and force of the driver's command. The control unit processes this information and generates the control signals for the wheel pressure modulators. Normally, the master brake cylinder is detached from the brake circuit. A pedal travel simulator creates normal pedal feedback. If ESP intervenes, the high-pressure reservoir supplies the required brake pressure quickly and precisely to selected wheels, without any driver involvement.

Electro-mechanical brake

In the future, the brakes system there will be no use of hydraulics and a brake-by-wire or electro-mechanical brake will be in production. The braking force will apply directly by electro-mechanical actuators. Brake cylinders, brake lines and hoses will have become components of the past.

The principle of Electro-mechanical brake is based on separating the hydraulic connection between the brake pedal and the wheel brake. An actuation unit replaces the conventional braking system. by depressing the brake pedal, which will consist of a pedal feel simulator and sensors, to pick up driver commands. Signals coming from this unit together with other sensors are transmitted electrically (by wire) to an electronic control module (ECM), which in turn conveys the sensed brake pressure to the wheel brakes. The functions of Electro-mechanical brake are compatible with all known systems such as ABS, ESP etc.

Contrary to the mechanical/hydraulic combination, the pedal no longer vibrates, thus lessening the risk of an inexperienced driver erroneously reducing brake pressure and terminating the ABS control mode. The electronic control unit computes the amount of braking deceleration desired by the driver from the sensed pedal travel and actuation force.

The electro-mechanical brake even goes one-step further and eliminates the brake cylinder, brake lines and hoses, as all these components are replaced by electric wiring. The use of electrics reduces maintenance expense and eliminates the expense of brake fluid disposal. Electro-mechanical brake measures the force of the driver's intention to brake the vehicle via sensors monitoring the system in the brake-pedal feel simulator. The ECU processes the signals received, links them where appropriate to data from other sensors and control systems, and calculates – in a split second the force to be generated by the brake caliper when pressing the brake pads.

The wheel brake modules essentially consist of an electric control unit, an electric motor and a transmission system. The electric motor and transmission system form the so-called actuator, which generates the brake application forces in the brake caliper. Each of the four actuators can deliver force up to several tons in milliseconds.

The Power requirements for the Electro-mechanical brake are great, and could overload the 12V system in the vehicle. Therefore, the electro-mechanical brake is design for a 42V system. An inverter/converter supply voltage to the Electro-mechanical brake and not from the vehicle's battery.

With no effort, the Electro-mechanical brake could include an electric parking brake function - to reduce the stopping distance.

This Brake system uses resolvers and encoder to provide continuous measurements of the absolute position and speed of the disc brake.

In contrast to encoders, the resolver provides two output signals so that it is always possible to find the absolute angular position. These types of system's using Position Sensor located on the brake pedal (function like a

Throttle Position Sensor), wheel speed, wheel angle sensor, tilt-side angel of the yaw rate, and acceleration sensors to determine the optimal pressure applied to each of the wheels.

This reduces the erosion of the front brake pads while reducing vehicle weight-transferred forward by the inertia force (inertia), which is produced when the brakes pressure applied.

Because these systems require optimally a high voltage, at least 42 volts, they are suitable to install in a hybrid electric vehicle.

Electronic parking brakes

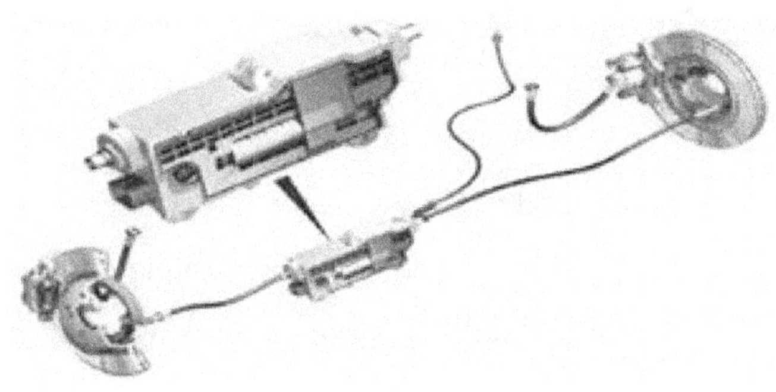

The new electric parking brake is installed as standard equipment. The conventional mechanical foot-operated parking brake with an additional pedal in the footwall is no longer present.

The electric parking brake operates via the electric parking brake switch. Pressing the electric parking brake switch engages the parking brake. To release the electric parking brake, you must pull the switch.

When the electric parking brake switch is pressed while the vehicle is stationary or moving at a speed less than 1 MPH, the duo servo brake (left and right drum brakes on the rear axle) closes regardless of the length of time the electric parking brake switch was pressed.

If the electric parking brake switch is applied to engage the parking brake while the vehicle is traveling less than 1 MPH an emergency braking sequence is initiated. The emergency brake function implemented by the traction system hydraulic unit, via all four wheels. ABS and ESP control may intervene if necessary.

The electric parking brake is released manually with the aid of a release tool. The transmission needs to be in "P".

Some manufacturers have developed braking systems-also by-wire electronic parking brake system to replace the system operated by a cable. Instead of pushing a foot pedal to set the parking brake, pressing a parking brake button activates it. Alternatively, you can set the parking brake operation automatically when the gear is transmitted to the P position. The advantages of this approach are:

The ability to activate the brakes automatically when the main brake system is fails. It can also combine with vehicle anti-theft system when the vehicle is parked, by locking the brakes.

Electronic throttle

Electronic throttle control has the following advantages over conventional cable: it eliminates the need for mechanical throttle cable, thus reducing the number of moving parts. These are replaced by a vehicle's computers by means of electronic systems that control the action of the engine except for the amount of incoming air. Eliminating the mechanical component of the cable and throttle, allows the system to respond faster, and therefore requires a minimum adjustment and maintenance.

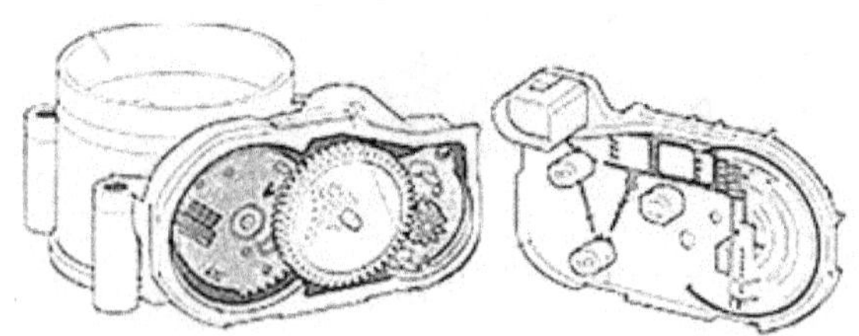

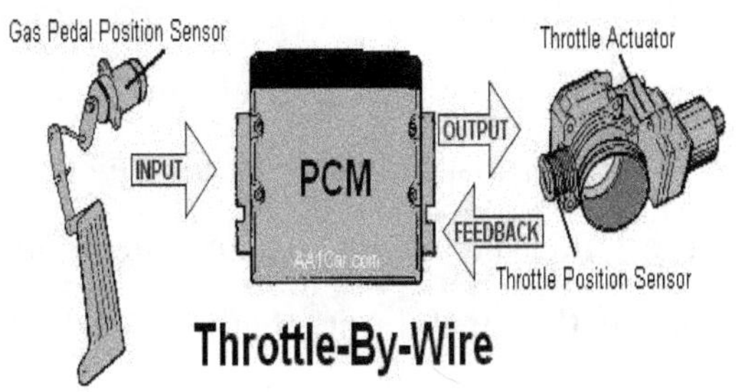

Acceleration system “by-wire" translates the movement of the accelerator pedal electrical signal. The movement of the pedal itself does not move in a similar way the throttle. If the driver tries to accelerate engine that could endanger the vehicle or the driver itself, the ECM would refrain from executing the command.

Maximum accuracy of the data improves the efficiency of the vehicle and fuel consumption. There are three basic components in electronic throttle control: two or three position sensors on the accelerator pedal, an electronically controlled throttle body with a small electric motor to open/close the throttle, and a control module (the ECM or a separate module).

On most vehicles, an electronic throttle fault will put the system into a "limp-in" mode that will limit engine speed. The throttle control system will remain in the limp-in mode until the fault will be diagnosed and repaired.

At the first sign of a problem, most electronic throttle controls are design to close the throttle and return to idle. for example, if the ECM detects a problem with a sensor, the system reverts to idle, preventing the throttle from opening.

Similarly, there are a number of terminations built into the system. For example, there are two sensors on the accelerator pedal to detect driver inputs. If a sensor malfunctions, or the two sensors report different readings, the system closes the throttle, idling the engine.

Most systems use a smart throttle motor. The throttle motor is the final gatekeeper that throttle signals need to go through before the throttle actually moves. If the throttle motor detects voltage or signals that did not come from the ECM, it is designed to turn the engine off. In some systems, which are already available from German manufacturers, the driver is allowed stepping, in to override the throttle system. If the throttle opens on its own, stepping on the brakes will close it.

The benefits of electronic throttle control are largely unnoticed by most drivers because the aim is to make the vehicle driving seamlessly. Electronic throttle control is also working 'behind the scenes' to improve the ease with which the driver can execute gear changes and deal with the dramatic torque changes associated with rapid accelerations and decelerations.

Active car safety features

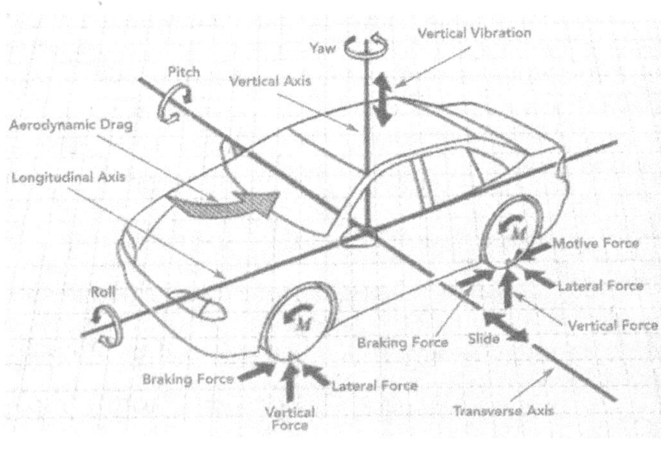

Understanding vehicle safety features is critical Enhanced technology is paving the way for more rigorous testing, and ultimately safer cars. Active safety features are those that help to prevent or mitigate road crashes. Unlike passive safety features - which designed to protect vehicle occupants once a crash has occurred - active safety features will engage to either prevent the crash from occurring, or reduce the severity of an unavoidable crash.

Intelligent Adaptation is any system that monitors vehicle speed and the local speed limit on a road and implements an action when the vehicle exceeds the speed limit. This done through an advisory system, through an intervention system where the driving systems of the vehicle are controlled automatically to reduce the vehicle's speed.

Intelligent speed adaptation uses information about the road on which the vehicle travels to make decisions about what the correct speed should be. . Many systems will provide information about locations where hazards may occur or where an enforcement action is indicated. This is particularly useful when drivers are in unfamiliar areas or when they pass through areas where variable speed limits are used.

There are several ways that active safety features activate a warning system when potentially dangerous situations are detected. Other safety features like electronic stability control, anti-lock braking systems, and brake assist monitor the vehicle's tires and brake systems for any signs that tailored braking is necessary in order to avoid a collision. Other active safety features work as failsafe measures to protect against driver errors. For example, brake override engages to deactivate the accelerator pedal in the event that both the gas and the brake pedal are pushed simultaneously.

Active safety features constantly monitor one or more aspects of the vehicle for potential hazards. These features work silently in the background, checking the rotation speed of the tires, the location of the vehicle within its lane, or the position of the gas and brake pedals relative to each other. When something problematic detected, active safety features act autonomously to correct the situation safely. Active safety features offer an extra layer of protection on the road.

Air bags

Most vehicles have airbags. Adaptive, or dual-stage front air bags, became standard. Crash sensors connected to an onboard computer detect a frontal collision and trigger the bags. The bags inflate in a few milliseconds- the blink of an eye—then immediately starts deflating.

Most air-bag systems now detect the presence, weight, and seat position for the driver and front passenger, and deactivate or de-power the front air bags as appropriate to minimize the chance of injury to drivers positioned close to the wheel, out-of-position occupants or children.

Side air bags protect in case of side-impact for front-seat passengers. Some automakers offer side bags for rear-seat passengers as well. The side air bags for the rear seat passenger are generally located in the side of the seat or behind the plastic trim. They help protect the torso, but most are not effective in protecting the head. Nearly all new models today also include additional side bags that deploy from above the windows and cover both front and rear side windows to prevent occupants from hitting their heads and to shield them from flying debris. A curtain bags often also stays 'inflated' longer in most cases to keep people from being ejected out of the vehicle during a rollover or a high-speed side crash. The better head-protection systems deploy the side-curtain bags if the system detects that the vehicle is beginning to roll over.

Antilock brakes (ABS)

Before antilock brakes came along, it was all too easy to lock up the wheels (stop them from turning) during hard braking. Sliding the front tires makes it impossible to steer, particularly on slippery surfaces. ABS prevents this from happening by using sensors at each wheel and a computer that maximizes braking action at each individual wheel to prevent lock-up. While braking the ABS system will allow the driver to retain steering control, so that the car can be maneuvered around an obstacle,

Traction control

Traction control systems are very common in vehicles today. The traction control system uses a computer to detect whether one (or more) of the wheels has begun to slip and lose traction.

As the ignition key is rotated through the phases of Off, Run, and Start, a system's bulb check is activated. The traction control system light should illuminate for 1 to 2 seconds and then go out as the vehicle starts and begins to run. If the light stays on, it means that the system is deactivated or there is a problem in the traction control system.

This electronically controlled system limits wheel spin during acceleration so that the drive wheels have maximum traction. It is particularly useful when starting in wet or icy conditions, and/or launching with a high-horsepower engine. Some traction-control systems can only operate at low speeds while others systems are active at all speed ranges.

Most traction-control systems use the car's antilock brake system to briefly constraint a spinning wheel. Some systems also may throttle back the engine, and up shift the transmission, to prevent wheel spin.

Loss of traction commonly occurs in either snow or ice when a moving wheel hits a patch of ice and begins to slip. When this lack of traction occurs, the traction control system shifts the power from the wheel that is slipping to the wheels that are still gripping. This transfer of power keeps the vehicle moving safely in the desired direction.

Many traction control systems will illuminate the warning light when the system detects a loss of traction, like in snowy or rainy weather. Typically, the light emit when the system intervenes to maintain traction.

If the traction control light illuminates and stays lit without blinking, this means that the traction control system has been deactivated and there is no traction control available.

Electronic stability control

Electronic stability control (ESC) takes traction control a step further. This system helps keep the vehicle on its intended path during a turn, to avoid sliding or skidding. It uses a computer linked to a series of sensors–detecting wheel speed, steering angle, sideways motion, and yaw (rotation). If the car drifts outside the driver's intended path, the stability-control system briefly brakes one or more wheels and, depending on the system, reduces engine power to pull the car back on course.

ESC is especially helpful with tall, top-heavy vehicles like sport-utilities and pickups, where it can also help keep the vehicle out of situations where it could roll over.

Vehicle Stability Control

Vehicle Stability Control (VSC) helps prevent wheels from slipping sideways when cornering or sudden steering. It also prevents side skids and stabilizes the vehicle while turning on a curve. According to NHTSA report, vehicles equipped with VSC compared to those without can effectively reduce single-vehicle accidents by 35% for automobiles and 67% for Sport Utility Vehicles (SUV).

The function of these systems is to restore the yaw rate of the vehicle as much as possible to the nominal motion expected, in spite of the road and driving conditions. Most of the ESP strategies act on the vehicle dynamics using the mechanical brakes to generate differential torques on the right and left wheels to produce a yaw moment.

When the vehicle senses a loss of traction or a slip, braking is applied automatically to all four individual wheels and engine power is reduced to help secure the safety of the vehicle. For example, if the steering wheel refuses to turn from over-speeding (under-steering), the vehicle will take control to steer toward the inner curve. When the vehicle spin from abrupt steering handling (over-steering), the vehicle will take control to steer toward the outer curve.

Vehicle Stability Control (VSC) designed to help the driver maintain vehicle control, but it is not a substitute for safe driving practices. In addition, the system will not be able to surpass the quality of the tires.

Hybrid electrical Vehicle

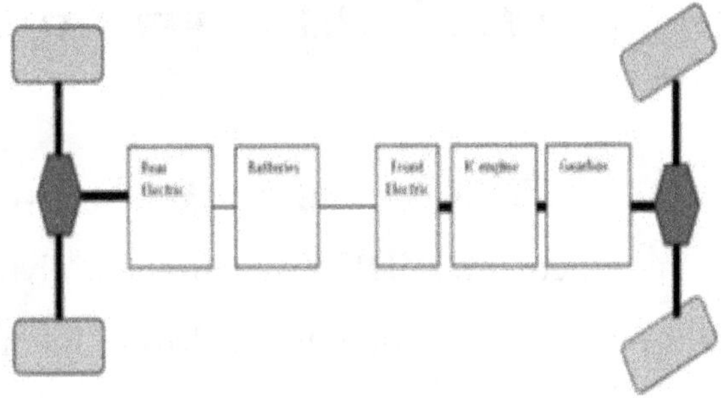

The VDIM system

VDIM integrated dynamic management. VDIM system is a dynamic system that can identify the driver's intentions accident situations and run various safety technologies to improve the overall performance of the vehicle in response to the demands of the driver.

VDIM System is special in two ways: First, it will start before the vehicle passes the limits of its operation, thereby achieving safety performance prophylactic high; second, few of the vehicle safety technologies (ABS, traction control, stability control, the ratio of steering changes and steering electrical power) combined smoothly in one system.

The VDIM system operates a variety of advanced safety features to prevent loss of gravity of the vehicle and loss of control. When a series of sensors detects the potential loss of gravity, the system uses a combination of single-wheel braking, turning and acceleration and steering assistance to help remedy the situation before a loss occurs.

The system is also equipped with traction control (TRAC) to prevent spinning while accelerating. A traction control system is typically a secondary function of the electronic stability control (ESP), designed to prevent loss of traction of driven road wheels. It is activated when throttle input and engine torque are mismatched to road surface conditions.

Yaw-rate sensor

A yaw-rate sensor is a gyroscopic device that measures a vehicle's angular velocity around its vertical axis. The angle between the vehicle's heading and vehicle actual movement direction is called slip angle, or skid angel, which is related to the yaw rate.

A Yaw Rate Sensor (or rotational speed sensor) measures a vehicle's angular velocity about its vertical axis in degrees or radians per second to determine the orientation of the vehicle as it hard-corners or threatens to rollover. In simpler terms, the yaw rate sensor is a key component in a vehicle's stability control or electronic stability control system. Yaw can be defined as the movement of an object turning on its vertical axis. The yaw rate sensor determines how far off-axis a car is "tilting" in a turn using gyroscopes to monitor the slip angle, the angle between the vehicle's heading and actual movement direction. The information is provided to the computer to re-calculate the wheel speed, steering angle and accelerator position. If the system senses too much yaw, the appropriate braking force will automatically applied.

By comparing the vehicle's actual yaw rate to the target yaw rate, the on-board computer can identify to what degree the vehicle may be under- or over-steering, and what corrective action, if any, is required.

Corrective action may include reducing engine power as well as applying the brake on one or more wheels to realign the vehicle.

The yaw rate sensor is typically located under the driver or passenger seat, mounted on the level floorboard to access the vehicle's center of gravity. After installation, a reset/recalibration procedure is generally required.

Lateral yaw sensor yaw -rate sensor monitors the vehicle's deviation from the vertical and the horizontal. Deviation represents the degree of stability of the vehicle. If the deviation angle reaches a certain threshold value, it indicates the form of vibration, and the third deviation occurred. Accordingly, the VSC control systems will be triggered. VSC control systems have two types of acceleration sensors: Vertical acceleration sensor and acceleration sensor horizontally. Both sensors provide the VSC control system the parameters of the vehicle.

When these sensors are not functioning, or after they have been replaced, the system should be re- calibrated to zero. Refer to the manufacturer information how to calibrate-zero and turn off the light problem. This function is available with a scanner, on the ABS list of options.

Front Camera

With cameras located at the center of the front grill and under the side mirrors, the driver can check hard-to-view areas by viewing the camera images on the screen. It provides the drivers a better view at intersections with unclear visibility and T-shaped junctions.

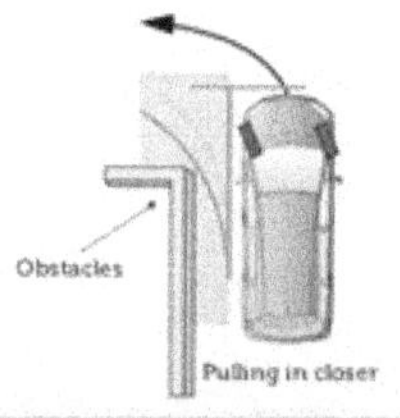

Toyota uses the built- in front prismatic camera in their vehicle and driver color screen to reduce the risk of a collision when approaching the intersection. The camera sends the image to the screen. When the vehicle comes to a crossroads where visibility is limited, and shows in view about 20 meters in both directions at 25 ° angle. It helps the driver see approaching vehicles, bicycles or pedestrians that could not otherwise visible. When the vehicle is moving backwards, back compact camera helps the driver see vehicles and obstacles behind the car, on the screen.

Smart Key

Hybrid electric vehicle implement smart key functions instead of a mechanical ignition switch, controlled by the vehicle control system. A smart key is an electronic access and authorization system, which become a standard in most cars.

The features of smart key include A wireless transmitter to lock/open the doors, and vehicle remote control for the alarm system. A key unit has metal key hidden inside to lock/unlock the driver's door when the smart key does not function.

The Vehicle operation has priority over all other ignition modes and it is accomplished by stepping on the brake pedal and pressing the ignition button once. When the key is inside the vehicle, the engine can be started by pressing the brake pedal and the Start/Stop button. If the Smart key is in the car and the lock button on the door handle is pressed, an alert will notify you that the key is inside and the car will not lock.

Smart Key can detect radio signals of other alarm system next to it. You cannot start the engine when the key-smart is outside the vehicle.

Pressing the ignition button once will supply power to start for 5 seconds. The computer can detect if there is no crankshaft sensor signal, (RPM) and Stops the starter operation.

If ignition takes longer, the ignition button should be pressed again, and you can extend it to 90 seconds at the most.

Door Locks

To unlock the doors using the available Smart Key, you have to have it with you, in your pocket, bag, or purse. The computer in the car will sense the key as you get closer and the courtesy lights inside will come on.

To unlock the door, slip your fingers behind one of the front doors and touch the inside handle's surface. The door will unlock automatically and you will hear two beeps.

To lock the doors, make sure they are all closed, and then touch the two grooves on top of the front door handle. You should hear a single beep.

If they do not lock and you hear a steady warning tone, one of the other doors is probably still ajar, or you left the Smart Key in the car.

To open the trunk, with the Smart Key, have the key with you and reach up under the trunk lid just above the license plate. You will feel a rectangular rubber button. Push it and the trunk will open.

The Smart Key also includes traditional remote lock, unlock, trunk open and emergency signal buttons.

If interference or a dead battery prevents the Smart Key from functioning properly, there is a mechanical key hidden in the fob itself.

You should know that strong radio signals, low batteries or even leaving the car parked and locked for two weeks or more could all affect the Smart Key system.

Key's Transponder

Transponder keys are devices designed to transmit a radio signal from a handheld device to a remote receiver. They are most commonly used to unlock and start vehicles, and keyless entry systems. Each one is programmed to start a specific vehicle, to reduce the possibility of theft. There are a number of ways to bypass transponders, however, most of which can be done with inexpensive equipment.

In vehicle-use transponder keys, the transmitter unit consists of a small microchip in the plastic part of a key. The microchip has a unique serial number, which is set during its initial programming. When a person wants to unlock or start a car, the chip sends a request to the car for it to validate the serial number and turn off the engine immobilizers. If the car does not recognize the number, it will remain locked and immobilized. A key that has not been programmed can still turn the engine over, but it will not be able to start it, since the immobilizers will still be activated.

The majority of keys to cars built after 1995 contain transponder chips. A transponder chip disarms a vehicle immobilizer when the car key is used to start the engine. Keyless transponder keys and remotes in all vehicles must be program to the same vehicle for them to function.

The transponders cannot be programmed in advance and come already set-up. Keyless entry remotes and the transponder keys are nearly always set up with the vehicle present. one of the few exceptions is if a transponder key is cloned from another key. Even then, it is advisable to test the key with the vehicle to ensure its operation.

Some vehicle manufacturers have built a system into their vehicles so that the vehicle owner performs self-program of a keyless entry remote, and/or transponder chip key. Some manufacturers did not build this system into their vehicles.

Some vehicle makers have the self-program system in some of their cars, and not in other models. Sometimes you will be able to self-program the keyless remote, but not the key.

Often a good automotive locksmith is a much better alternative to a dealer if you require assistance in setting up keyless remotes & chipped keys.

Programming of keyless entry remote fobs can take patience and persistence. Please read all the information few times so you can understand the procedures.

Add/Replace car keys

When you add a key, you must have all the keys handy before beginning the process. You have to use a key that already was programmed before. It is imperative to note that most places are taking advantage of the remote control frequency of 433 MHz (MHz).

Transponder and their chip can be replaced separately or together. The Encoding card can be replaced with a card encoding used or new with the original key software. When replacing transponder with Transponder used, you have at least one original key that belongs to transform he will learn the key Master- code once.

To start, you must first 'Add' the old transponder. The key Master-code is now ready to be duplicated.

The scanner will ask if you want to skip programing the old keys.

If you answer yes, you want to re-use the existing keys. If you answer no, you want to program the new key. The second phase of the process is to 'add.'

The master key code will be copied so that it could be replaced. When adding new transponder it makes no difference if it is new, because the transponder and chip being replaced, so use all the keys.

It is significant to note that the reason the new keys needed is that they are not compatible with the master code of the original keys.

Safety rules - high voltage:

What you should know:

Voltage over 60 volts and 1/2 amps can kill a human.

- High voltage is deadly and dangerous than direct current alternating current.
- Always use appropriate isolating gloves before each operation with high voltage.
- Do not make repairs or service by the unprofessional employees even if it look simple.
- Do not use pneumatic or electrical equipment.
- Do not handle wet electrical components, or when your hands or feet are wet.
- In case of an incident, remove the main battery breaker key of the HV battery to cut off the electricity immediately.
- The manufacturer's instructions should be read and you must perform accordingly.
- Do not push the vehicle over 4 MPH or tow the car when the driving wheels on the

road.

- When removing high voltage component place it on an insulated surface.

Digital voltmeter CAT- III (DMM) typically rated 1000V, is required to perform measurements with high voltage system of electric vehicles.

After the removal of an HV battery, place it on a rubber-isolated table.

When lifting the vehicle at the repair shop facility, pay attention not to put the lift's arms near the orange HV cables, under the vehicle.

Main service plug

Safety system for the HV battery contains a master service plug with high voltage fuse inside it. When the service plug is removed, it cuts the high voltage circuit and turns off the relay of the SMR relays.

For safety reasons, you must always switch the ignition to OFF before removing the service plug. When it is necessary for a technician to work with the high voltage, disconnect the main service plug. Wait at least 5 minutes. For Honda vehicle, its 25 minutes. Perform a voltage checks the orange cable with voltmeter and if the voltage is 12 volts or less, the vehicle is safe for service.

The HV battery should be removed from the vehicle while charging it.

Remove any interior or access panels from the top of the HV battery pack. Put on your special gloves. Disabling the battery pack is the first step to work with any hybrid system. While the full battery voltage, anywhere from 144 to nearly 300 volts DC, is available at the output terminals, battery manufacturers have left us a simple way to deactivate the contacts, and safely remove the battery pack. The HV battery pack is a long series of individual cells. Somewhere near the electrical center of this string is a service switch or removable plug.

Now you can unlock the clamp of the service plug and remove the plug from its socket. You have just disconnected the battery pack internally, halfway of the battery cells pack. This means the voltage output measure at the connectors of the HV battery is zero. You can remove the battery now.

When you are done, simply put your gloves back on and insert the service plug back into its socket. Make sure the plug is fully seated. Now you can reinstall any access panels or trim and reconnect the 12-volt battery's negative clamp. You will need to key in any radio security code and the station presets.

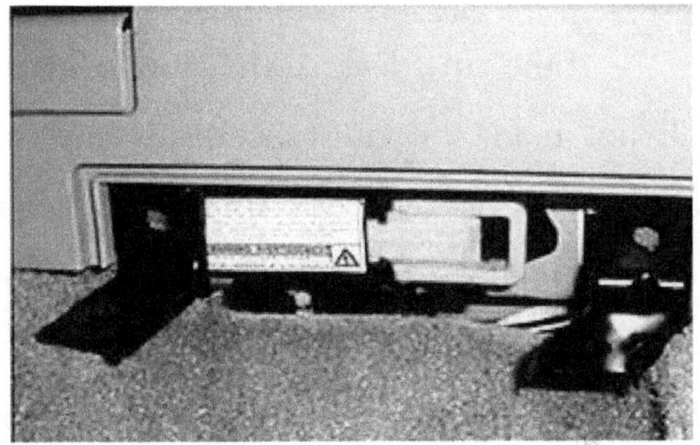

Rubber gloves

Before working with high voltage, be sure the Linesman high voltage gloves are handy. Ensure that gloves are rated for at least 1,000 volts and have a '0' readiness. You should test the rubber gloves using the following procedure:

Turn over the glove from the open section to the bottom of the glove. Because of the applied pressure, the air is held in the glove.

If the air is leaking the glove's fingers will not hold and must be disposed. You must test the gloves every six months by an accredited laboratory. The gloves need to be tested carefully before each use. If the gloves show signs of wear & tear, they cannot be used.

Fire in Hybrid vehicle

There is no unusual hazards from regular hybrid electric vehicle involved in the accident and caught fire but must exercise caution in extracting extensive and ongoing, especially when cutting required for extracting vehicle. Switch off the ignition switch and disconnect the main circuit breaker plug, to prevent contacting with high voltage cables. The cables are usually orange, although there are some blue cables to indicate a medium voltage system. Cables may be covered with black plastic and not be visible.

Almost all electric cars currently on their way to the marketplace use large arrays of lithium-ion (Li-ion) batteries. Li-ion batteries have a tendency to catch fire and occasionally explode.

The amount of energy packed into each battery is quite large relative to its size. That is why Li-ion batteries have become the battery of choice for electric cars, but it also gives them some of the characteristics of an explosive. Moreover, if the small Li-ion batteries in cell phones are a danger, imagine how much more dangerous the huge battery array in a car can be. The battery pack in a Tesla Roadster, for instance, consists of 6,500-7500 batteries under the hood of a single car.

Hybrid vehicle submerged in water

There is no danger of electric shock from touching parts of the car in the water or out of water. Remove the vehicle from the water and then use one of the procedures described below to prevent the flow of electricity through high voltage cables.

The reader should not assume that if the engine is off, it is also disabled. The vehicle may be powered by the electric motor only. Be sure the 'READY' light stays off.

When an electric hybrid electric vehicle involved in an accident or has been in deep water, cable insulation should be tested before the high voltage system is turned back on. Insulation Tester, often called Megger, such fluke-1587, is used to check the insulation. Acceptable resistance readings are more than one million ohms. If the reading is less than one million ohms, it is necessary to perform additional tests to locate the isolation break.

Towing

Towing with front wheel drive (or any dive-wheels) on the ground can cause engine to produce electricity. Insulation power cables can leak and cause a fire. When vehicles are towed, the driving wheels must be lifted off ground. For the same reason, do not push the car faster than 3MPH.

After removing the main service plug, do not press the ignition button to READY position, unless the manufacturer's instructions indicate it. This may cause a malfunction in the system. When removing or installing a converter, do not touch the inverter terminals so to prevent static damage.

Troubleshooting

The first generation of hybrid car engines should provide a lot of information concerning the types of problems particular to hybrid cars.

Diagnosing a Hybrid system is much similar to that on other late model vehicles. If a problem occurs with something that is not part of the hybrid system, it will set diagnostic code that corresponds to that fault, and turn on the Check Engine light. However, if a fault involves any of the components in the hybrid system (no start, stalls or battery problems etc.), diagnostics will likely require a Dealer scan tool (or aftermarket scan tool, with similar software capabilities.

To access the onboard diagnostics you need a scan tool. Most of the inexpensive scanners will display codes only and will not provide detailed sensor values or other important data. Most vehicle manufacturers use both generic OBD II codes and their own factory OEM diagnostic codes. Almost any scan tool these days will read the generic OBD II codes, but it often requires more expensive scan tool or scanner software to read all the factory codes. Without this information, your diagnostic capabilities will be extremely limited because the factory codes cover faults that may occur in the hybrid system.

If the Malfunction Indicator Lamp (MIL) is on, you will want to plug in your scan tool to read the codes and any freeze frame data that might have been captured to help you in your diagnosis. Depending on what the code you get, look at related PIDs to see the sensors performance and the way the system operates. Problem areas that are going to require some hybrid knowledge include any faults related to the high voltage battery or charging system, any faults related to engine cranking, transmission, or the hybrid control system.

Manufacturers provide detailed diagnostic charts for all fault codes. The charts include the fault description, which components are involved, the wiring diagrams and the procedure needed to isolate the fault.

Therefore, you need to read service bulletins and be familiar with the various hybrid components and have a scan tool that can read the proper PIDs and freeze frame data.

Because the inverter/converter is part of the high voltage system, you do not want to touch anything until the HV (high voltage) battery you disconnect it by pulling out the power service plug in the trunk. You then must wait at least five minutes for the high voltage capacitors in the inverter/converter to discharge before proceeding with repairs to the cooling system or inverter/converter assembly.

If a 2004 Prius suddenly shuts down during rainy weather, the problem may be water leaking past the hood cowl seal. If water drips on the ignition system circuits, it can cause a misfire. The hybrid software then shuts down the engine and the car will not drive. Dry the ignition system with air and repair the leaky hood seal.

a faulty 12-volt coolant pump in a Prius is caused by the Inverter/converter cooling failures.

Battery cooling problems can cause premature battery failure and turn on the MIL lamp.

When you see codes C1341, C1342, C1343, and C1344 after replacing the brake pads clear the codes with a scanner and then test-drive the vehicle.

Code P3191 appears when the engine oil viscosity is too high and causes resistance.

Excessive filling of engine oil increases the resistance, increases fuel consumption, and no Starting codes. You cannot start the engine in this mode. Fix the oil leaks, add oil as needed, and clear the codes with a scanner.

The first step in the service of the cooling system is to disable the storage tank pump - by disconnecting the gray dual-wire connector.

Wearing eye protection glasses, empty the coolant by opening the three drain plugs, fill the system with coolant, and connect the scanner to operate the pump.

Turn the ignition on by pressing the ignition switch button twice, and turn-on the water storage pump. You have to add coolant until there is no air in the system, and the pump operation becomes silent and causes a liquid to spin in the reservoir tank.

* * * * *

DTC code C1515 does not indicate a problem. This code is set when the steering torque sensor is not calibrated. Calibrate the torque sensor to zero readiness, and then clear the code.

Code C1516 does not indicate a problem. The code appears when the process of calibrating the torque sensor was not performed properly. Try to re-calibrate it again and clear the code.

Temporary's failures with no specific code can be diagnosed with a scanner. In EPS menu, select RECORDS to display a list of information related to the engine overheating or EPS open circuit. It will not have any related code.

If PRIUS 2004 and newer suddenly shut down during rainy weather, there is a water leak into the engine compartment through rubber-insulated hood.

P 3009 code appears when there is a high voltage leak in the high voltage system; use the Mugger Checker to measure the insulation resistance between the cable and the vehicle. If Mugger tester is not available, try this procedure to isolate the problem: using a scanner, clear the codes from the computer press the ignition button to Ready and check to see if the code still exists. If there is a code, unplug the HV (high voltage) battery and re-connect again. If the code is back, then the problem in HV (high voltage) battery or the cable connections.

P0AA6 code or P3009, with information codes 526 and 614-you need to replace the converter system.

When the Ready ignition button is not lit up and/or have codes P3140, P3141, or P0A0D -make sure to slide the main utility plug lever into place.

When the codes are P3130, P3125, which is electrical power problems, or the Ready ignition button is inoperative, the auxiliary pump of converter's cooling system does not operate. When code P3125 exist with no other code it denote the AC/DC inverter malfunctions, and you need to check the ECM.

Code P3009 indicates a high voltage leakage from the high voltage battery to vehicle frame.

Continuous operation of ICE

when you need the Internal combustion engine to run continuously, without stopping for testing and diagnosis, start the test mode within 60 seconds:

▪ Keep your foot off the brake pedals. Then press the ignition button twice.

▪ Next step, press the accelerator pedal to the floor and release it -twice.

▪ press the brake pedal and keep it so with your left foot throughout this process.

▪ move the gear shifter to neutral position then press the accelerator pedal to the floor and release it - twice.

▪ move the gearshift back to P (park) and then press the accelerator pedal to the floor and release it - twice.

▪ The HV (high voltage) light on the

instrument panel should blink.

- press the ignition button. The engine should start and run at 1000RPM.
- press the ignition button once more to Stop the engine. At this point, turn off the display information, and that is the last phase of the test mode.

The important thing to remember: Never drive the vehicle while the engine is in test mode!

End

www.ingramcontent.com/pod-product-compliance
Lightning Source LLC
Chambersburg PA
CBHW051855170526
45168CB00001B/116

* 9 7 8 1 5 1 2 3 2 5 9 6 6 *